SpringerBriefs in Physics

Series Editors

Balasubramanian Ananthanarayan, Centre for High Energy Physics, Indian Institute of Science, Bangalore, India

Egor Babaev, Department of Physics, Royal Institute of Technology, Stockholm, Sweden

Malcolm Bremer, H. H. Wills Physics Laboratory, University of Bristol, Bristol, UK

Xavier Calmet, Department of Physics and Astronomy, University of Sussex, Brighton, UK

Francesca Di Lodovico, Department of Physics, Queen Mary University of London, London, UK

Pablo D. Esquinazi, Institute for Experimental Physics II, University of Leipzig, Leipzig, Germany

Maarten Hoogerland, University of Auckland, Auckland, New Zealand

Eric Le Ru, School of Chemical and Physical Sciences, Victoria University of Wellington, Kelburn, New Zealand

Dario Narducci, University of Milano-Bicocca, Milan, Italy

James Overduin, Towson University, Towson, USA

Vesselin Petkov, Montreal, Canada

Stefan Theisen, Max-Planck-Institut für Gravitationsphysik, Golm, Germany

Charles H. T. Wang, Department of Physics, University of Aberdeen, Aberdeen, UK

James D. Wells, Department of Physics, University of Michigan, Ann Arbor, MI, USA

Andrew Whitaker, Department of Physics and Astronomy, Queen's University Belfast, Belfast, UK

SpringerBriefs in Physics are a series of slim high-quality publications encompassing the entire spectrum of physics. Manuscripts for SpringerBriefs in Physics will be evaluated by Springer and by members of the Editorial Board. Proposals and other communication should be sent to your Publishing Editors at Springer.

Featuring compact volumes of 50 to 125 pages (approximately 20,000–45,000 words), Briefs are shorter than a conventional book but longer than a journal article. Thus, Briefs serve as timely, concise tools for students, researchers, and professionals.

Typical texts for publication might include:

- A snapshot review of the current state of a hot or emerging field
- A concise introduction to core concepts that students must understand in order to make independent contributions
- An extended research report giving more details and discussion than is possible in a conventional journal article
- A manual describing underlying principles and best practices for an experimental technique
- An essay exploring new ideas within physics, related philosophical issues, or broader topics such as science and society

Briefs allow authors to present their ideas and readers to absorb them with minimal time investment. Briefs will be published as part of Springer's eBook collection, with millions of users worldwide. In addition, they will be available, just like other books, for individual print and electronic purchase. Briefs are characterized by fast, global electronic dissemination, straightforward publishing agreements, easy-to-use manuscript preparation and formatting guidelines, and expedited production schedules. We aim for publication 8–12 weeks after acceptance.

Jaroslav Zamastil • Tereza Uhlířová

An Algebraic Approach to the Many-Electron Problem

 Springer

Jaroslav Zamastil
Dept of Chemical Physics & Optics
Charles University
Praha 2, Czech Republic

Tereza Uhlířová
Dept of Chemical Physics & Optics
Charles University
Praha 2, Czech Republic

ISSN 2191-5423   ISSN 2191-5431   (electronic)
SpringerBriefs in Physics
ISBN 978-3-031-87827-5   ISBN 978-3-031-87825-1   (eBook)
https://doi.org/10.1007/978-3-031-87825-1

© The Editor(s) (if applicable) and The Author(s), under exclusive license to Springer Nature Switzerland AG 2025

This work is subject to copyright. All rights are solely and exclusively licensed by the Publisher, whether the whole or part of the material is concerned, specifically the rights of translation, reprinting, reuse of illustrations, recitation, broadcasting, reproduction on microfilms or in any other physical way, and transmission or information storage and retrieval, electronic adaptation, computer software, or by similar or dissimilar methodology now known or hereafter developed.
The use of general descriptive names, registered names, trademarks, service marks, etc. in this publication does not imply, even in the absence of a specific statement, that such names are exempt from the relevant protective laws and regulations and therefore free for general use.
The publisher, the authors and the editors are safe to assume that the advice and information in this book are believed to be true and accurate at the date of publication. Neither the publisher nor the authors or the editors give a warranty, expressed or implied, with respect to the material contained herein or for any errors or omissions that may have been made. The publisher remains neutral with regard to jurisdictional claims in published maps and institutional affiliations.

This Springer imprint is published by the registered company Springer Nature Switzerland AG
The registered company address is: Gewerbestrasse 11, 6330 Cham, Switzerland

If disposing of this product, please recycle the paper.

Preface

The many-electron problem in quantum mechanics is the basis for our understanding of the atomic and molecular structure, the nature of the chemical bond, the rates and mechanisms of chemical reactions, and so on. The Hartree-Fock (HF) method reduces the many-electron problem to the problem of one electron moving in an effective field of nuclei and other electrons. Although the HF method explains semi-quantitatively the main features of the problem, for instance, it is able to explain Bohr's Aufbau Principle for filling electron shells in the atoms, it does not suffice for a quantitative comparison with experiment. That is, the HF method does not provide a predictive, and whence useful, theory. To obtain such a theory, one has to take into account what is called the dynamical correlation between electrons. This brings us to the realm of so-called post-HF approaches. Among those, the coupled-cluster (cc) method plays a prominent role due to its correct scaling behavior with increasing the number of electrons, the so-called size-extensivity.

This book grew out of the author's dissatisfaction with the usual presentation of cc method. There are two common ways to present it. The first route, see for instance [1] [1], is to stay within the framework of ordinary quantum mechanics. This means to use Slater determinants to fulfill the requirement of the antisymmetry of the many-electron wave function. Proceeding in this way, one encounters very early rather formidable combinatorics [2]. So at some point, the authors stop and only indicate a general strategy how to proceed. The second route is to use field methods. The authors derive the rules for construction of Feynman-like diagrams and their evaluation. This was the original presentation of the cc method by its inventor Jiří Čížek [3]. This route is followed for instance in [4, 5, 6, 7]. We do not find this diagrammatic approach illuminating either, but this is clearly a matter of taste. In this book, we follow the third route, pioneered by Josef Paldus [8]. We use field methods, but we proceed in an algebraic, not diagrammatic, manner. In our view, and also in our teaching experience, this is the most simple and understandable way

---

[1] We strongly recommend this excellent textbook to reader's attention.

to derive and use the cc method. Once one absorbs the formalism of a quantized electron field, everything else is straightforward.

As it is customary in most ab initio calculations and as was alluded to above, we start from the independent particle model, the HF method. We assume throughout the most of the book that the HF solution exists and is unique. This covers nearly all neutral closed-shell molecules not very far from equilibrium geometry. Thus we restrict ourselves to what is considered as standard and firmly established. The situations where this is not the case are certainly interesting from both the methodological and practical point of view, but they are outside the scope of this introductory treatment. For a not so too out-of-date state of the affairs and a guide to literature we refer reader to [4]. For a fascinating account of the history of the cc method, see Josef Paldus's essay [9].

This book is organized as follows. In Chap. 1 we introduce the notion of the quantized electron field and show how the N-electron Hamiltonian can be expressed in its language. In the following Chap. 2 we introduce the notion of the Fermi vacuum and derive the Hartree-Fock equations and conditions for the stability of their solutions. These two chapters are preparatory for Chap. 3. There, we first discuss the so-called method of configuration interaction, which is commonly used for accounting for dynamical correlation between electrons. We point out the size-extensivity problem and show how this problem is solved within the cc approach. We then proceed to derive the cc equations in spin-orbital form. Chapter 4 mostly deals with practical aspects of the cc method. We show how one can take advantage of the permutational and spin symmetries, and how to practically solve cc equations. We illustrate this whole approach on the Hubbard model of benzene, the simplest quasi-realistic model of electron correlation. Finally, we briefly discuss the use of the cc method for one-electron open-shell systems.

Prerequisites

Knowledge of quantum mechanics on the level of standard courses, see, e.g., [10, 11, 12, 13], is assumed.

Acknowledgments JZ would like to express his gratitude to Jiří Čížek (1938–2024) and Josef Paldus (1935–2023) for their long and lasting friendship. Additionally, thanks are owed to our colleague Vojtěch Patkóš for carefully reading the manuscript and pointing to us its several shortcomings.

Praha 2, Czech Republic
Jaroslav Zamastil
Tereza Uhlířová

Contents

1	**Quantized Electron Field**	1
1.1	Many-electron Problem	1
1.2	Quantization of Electron Field	2
1.3	The One- and Two-particle Operators	7
2	**Hartree-Fock Approximation**	11
2.1	Energy of the N-particle State	11
2.2	Fermi Vacuum	13
2.3	Hartree-Fock Equations	16
2.4	Spin-restricted Form of Hartree-Fock Equations	18
2.5	Stability Conditions	19
2.6	Spin Adapted Stability Matrix	22
3	**Coupled Cluster Method**	29
3.1	Configuration Interaction	29
3.2	Problem of Size Extensivity	32
3.3	Coupled Cluster Equations in Matrix Form	34
3.4	Coupled Cluster Equations in Spin-orbital Form	38
4	**Further Developments**	43
4.1	Adaptation to Permutational Symmetry	43
4.2	Adaptation to Spin Symmetry	45
4.3	Perturbative Solution of Coupled Cluster Equations	49
4.4	Iterative Solution of Coupled Cluster Equations	50
4.5	Hubbard Model of Benzene	51
4.6	Inclusion of Monoexcitations	55
4.7	Perturbative Inclusion of Triexcitations	59
4.8	One-electron Open Shells	62
	4.8.1 Combination of Coupled Clusters and Configuration Interaction	62

4.8.2	Method for Obtaining Bound-state Energies	64
4.8.3	Matrix Elements of Configuration Interaction	65
4.8.4	Perturbative Inclusion of Five-particle States	67
4.8.5	Symmetry Adaptation	68

References 71

Chapter 1
Quantized Electron Field

In this chapter we introduce the notion of the quantized electron field and show how the N-electron Hamiltonian can be expressed in its language.

1.1 Many-electron Problem

The Hamiltonian of the N-electron problem has a generic form

$$\hat{H} = \sum_{i=1}^{N} \hat{z}(i) + \sum_{i=1}^{N} \sum_{j=i+1}^{N} \hat{v}(i,j), \tag{1.1}$$

where $\hat{z}$ and $\hat{v}$ are referred to as one- and two-particle operators, respectively. For example, in molecules, one employs the so-called Born-Oppenheimer approximation [1]: electrons move in the field of fixed nuclei. Considering, for instance, a diatomic molecule, the operators $\hat{z}$ and $\hat{v}$ take the form in atomic units

$$\hat{z}(i) = -\frac{1}{2}\nabla^2_{(i)} - \frac{Z_A}{r_{iA}} - \frac{Z_B}{r_{iB}} \tag{1.2}$$

and

$$\hat{v}(i,j) = \frac{1}{r_{ij}}, \tag{1.3}$$

respectively. Here, Z_A and Z_B are the charges of the nuclei A and B in the units of the elementary charge, in this units the charge of the electron is minus one, r_{iA} and r_{iB} are the distances between the i-th electron and the nuclei A and B and r_{ij} is the distance between i-th and j-th electron. One is interested in the solution of the stationary Schrödinger equation

$$\hat{H}|\psi\rangle = E|\psi\rangle. \quad (1.4)$$

Once the solution of this equation is known, one can determine experimental quantities such as the positions and intensities of spectral lines and so on.

The solution of Eq. (1.4) is a formidable task indeed. One has to resort to different levels of approximations and numerical computations. Nonetheless, on any level of approximation, one wants, at least in truly *ab initio* calculation, the approximate wave function to share the symmetry properties of the exact wave function. In the most general case, there are no symmetries, except for two: the permutational and spin symmetries. The permutational symmetry means the exact electron wave function has to be antisymmetric with respect to exchange of any pair of electrons. It is this requirement that makes the field approach more advantageous than the particle one.

To illustrate the point, let us first consider the two electron case. The simplest choice of a two-electron wave function which is antisymmetric with respect to the interchange of the first and second electrons is the so-called *Slater determinant*

$$|\psi\rangle = \frac{1}{\sqrt{2}} \begin{vmatrix} |a\rangle_1 & |b\rangle_1 \\ |a\rangle_2 & |b\rangle_2 \end{vmatrix} = \frac{1}{\sqrt{2}} \left(|a\rangle_1 |b\rangle_2 - |b\rangle_1 |a\rangle_2 \right), \quad (1.5)$$

where $|a\rangle$ and $|b\rangle$ design spin-orbitals and the subscripts refer to the electrons. The factor $1/\sqrt{2}$ serves to guarantee the proper normalization $\langle\psi|\psi\rangle = 1$, since the spin-orbitals are assumed to be orthonormal: $\langle a|a\rangle = \langle b|b\rangle = 1$ and $\langle a|b\rangle = 0$.

Next consider the three-electron case. The simplest three-electron wave function antisymmetric with respect to interchange of any pair of electrons is the Slater determinant

$$|\psi\rangle = \frac{1}{\sqrt{3!}} \begin{vmatrix} |a\rangle_1 & |b\rangle_1 & |c\rangle_1 \\ |a\rangle_2 & |b\rangle_2 & |c\rangle_2 \\ |a\rangle_3 & |b\rangle_3 & |c\rangle_3 \end{vmatrix}. \quad (1.6)$$

Clearly, going to the N-electrons, the Slater determinant comprises $N!$ terms. Evaluating the matrix elements of the Hamiltonian (1.1) between the two Slater determinants could be horrible combinatorial task.

1.2 Quantization of Electron Field

According to our present view of the Nature, electrons are quanta of an electron-positron field. When an electron collides with a positron, they could disappear with a flash of two photons—quanta of electromagnetic field. In the nuclear beta decay,

1.2 Quantization of Electron Field

electrons are created at the very moment of neutron disintegration together with protons and antineutrinos. However, here we are interested in the determination of the stationary solution of the Schrödinger equation, Eq. (1.4), for atoms, molecules, etc., where we have N electrons that are there, so to speak, for ever. Thus, the view that electrons are particles that cannot be created or annihilated, is perfectly adequate. Nonetheless, as we shall see, even in this, strictly non-relativistic regime, the view of electrons as excitations of the underlying electron field, is superior to the view of electrons as forever existing particles.

We quantize the electron field in the same spirit as any other field.[1]

1. We regard the Schrödinger equation for one electron

$$\left(i\frac{\partial}{\partial t} - \hat{h}\right)\psi(\boldsymbol{r},t) = 0, \tag{1.7}$$

where $\hat{h}$ is a one-particle Hamiltonian, as an equation for the classical electron field in the same way the Maxwell equations are regarded as equations for the classical electromagnetic field. This was, in fact, the original Schrödinger's view. Note, though, that there is a double duplicity here. First, the electron has an internal degree of freedom, the spin, so ψ is the spinor wave function with two components. Second, the wave function ψ is clearly complex, so it has the real and imaginary parts. Thus, instead of decomposing the wave function ψ into its real and imaginary parts, we can regard the Hermitian conjugate field ψ^+ as independent. Thus, we have the second equation

$$\psi^+(\boldsymbol{r},t)\left(-i\frac{\partial}{\partial t} - \hat{h}\right) = 0. \tag{1.8}$$

Note that the following holds irrespective of the precise form of the one-particle Hamiltonian $\hat{h}$.

2. We expand the electron field into suitable modes

$$\psi(\boldsymbol{r},t) = \sum_\sigma b_\sigma(t) U_\sigma(\boldsymbol{r}). \tag{1.9}$$

The modes, called *spin-orbitals*, are the eigenstates of the one-particle Hamiltonian $\hat{h}$

$$\hat{h} U_\sigma(\boldsymbol{r}) = \epsilon_\sigma U_\sigma(\boldsymbol{r}). \tag{1.10}$$

---

[1] The following exposition follows closely that given in [14].

The Hamiltonian $\hat{h}$ is Hermitian, so its eigenstates form a complete and orthonormal one-particle basis,

$$\sum_\sigma |U_\sigma\rangle\langle U_\sigma| = 1 \qquad (1.11)$$

and

$$\langle U_\sigma | U_\rho\rangle = \delta_\rho^\sigma, \qquad (1.12)$$

respectively. Equation (1.9) is obtained by projecting Eq. (1.11) on the state $|\psi\rangle$ and setting $b_\sigma(t) = \langle U_\sigma | \psi(t)\rangle$. Taking the Hermitian conjugate of Eq. (1.9), we get

$$\psi^+(\boldsymbol{r}, t) = \sum_\sigma b_\sigma^+(t) U_\sigma^+(\boldsymbol{r}). \qquad (1.13)$$

3. After inserting the decomposition (1.9) into the Schrödinger equation (1.7) and using the orthonormality relations, Eq. (1.12), we obtain the evolution equations for the coefficients b_σ

$$\left(i\frac{d}{dt} - \epsilon_\sigma\right) b_\sigma(t) = 0. \qquad (1.14)$$

Likewise, by substitution Eq. (1.13) into Eq. (1.8) we obtain

$$\left(-i\frac{d}{dt} - \epsilon_\sigma\right) b_\sigma^+(t) = 0. \qquad (1.15)$$

The last equation can also be obtained by taking Hermitian conjugate of Eq. (1.14).

4. Equations (1.14) and (1.15) can be regarded as the evolution equation of the classical theory. They can be derived from the Hamiltonian

$$H_e = \sum_\sigma \epsilon_\sigma b_\sigma^+ b_\sigma \qquad (1.16)$$

and the Hamilton equations of motion $\dot{q}_j = \frac{\partial H_e}{\partial p_j}$, $\dot{p}_j = -\frac{\partial H_e}{\partial q_j}$ if b_σ and ib_σ^+ are regarded as canonical coordinates and momenta, respectively

$$\dot{b}_\sigma = \frac{\partial H_e}{\partial ib_\sigma^+} = -i\epsilon_\sigma b_\sigma, \quad i\dot{b}_\sigma^+ = -\frac{\partial H_e}{\partial b_\sigma} = -\epsilon_\sigma b_\sigma^+.$$

1.2 Quantization of Electron Field

5. We impose the canonical *anti*commutation relations on the canonical coordinates and momenta

$$\{\hat{b}_\sigma, i\hat{b}_\rho^+\} = i\delta_\sigma^\rho. \tag{1.17}$$

Further, we stipulate the canonical coordinates to be mutually anticommuting

$$\{\hat{b}_\sigma, \hat{b}_\rho\} = 0. \tag{1.18}$$

Taking the Hermitian conjugate of the last equation we find that also the canonical momenta are anticommuting

$$\{\hat{b}_\sigma^+, \hat{b}_\rho^+\} = 0. \tag{1.19}$$

We will comment on the last three relations shortly.

6. The vacuum state of the electron field is determined by the equation

$$\hat{b}_\sigma |0\rangle = 0 \tag{1.20}$$

valid for all modes σ. Taking the Hermitian conjugate of the last equation, we obtain

$$\langle 0| \hat{b}_\sigma^+ = 0. \tag{1.21}$$

7. The quantum Hamiltonian is obtained by inserting the corresponding operators for canonical coordinates and momenta into the classical Hamiltonian

$$\hat{H}_e = \sum_\sigma \epsilon_\sigma \hat{b}_\sigma^+ \hat{b}_\sigma. \tag{1.22}$$

Likewise, going back to Eqs. (1.9) and (1.13) and substituting for the canonical coordinates and momenta, $b(t)$ and $ib^+(t)$, the corresponding quantum operators $\hat{b}$ and $i\hat{b}^+$, respectively, we obtain the decomposition of the electron field and its conjugate that we will use extensively in the following

$$\hat{\psi}(r) = \sum_\sigma \hat{b}_\sigma U_\sigma(r), \quad \hat{\psi}^+(r) = \sum_\sigma \hat{b}_\sigma^+ U_\sigma^+(r). \tag{1.23}$$

Equations (1.17)–(1.19) provide the interpretation of the operators $\hat{b}^+$ and $\hat{b}$ as creation and annihilation operators of the electrons. To see this, we evaluate first the action of the Hamiltonian on the vacuum state

$$\hat{H}_e |0\rangle = 0|0\rangle, \tag{1.24}$$

by virtue of Eq. (1.20). Next consider the action of the Hamiltonian on the state $\hat{b}_a^+|0\rangle$:

$$\hat{H}_e \hat{b}_a^+|0\rangle = \sum_\sigma \epsilon_\sigma \hat{b}_\sigma^+ \hat{b}_\sigma \hat{b}_a^+|0\rangle = \quad (1.25)$$

$$= \sum_\sigma \epsilon_\sigma \hat{b}_\sigma^+ \left(-\hat{b}_a^+ \hat{b}_\sigma + \delta_\sigma^a\right)|0\rangle = \epsilon_a \hat{b}_a^+|0\rangle \,.$$

The first equality follows from Eq. (1.22), the second one from Eq. (1.17) and the last one from Eq. (1.20). The last two equations tell us that the states $|0\rangle$ and $\hat{b}_a^+|0\rangle$ are the eigenstates of the Hamiltonian (1.22) with the energies 0 and ϵ_a, respectively. Whence we interpret the operator $\hat{b}_a^+$ as the creation operator of the electron in the spin-orbital $|a\rangle$. If we act on the state $\hat{b}_a^+|0\rangle$ with the operator $\hat{b}_a$, we get back to the vacuum state

$$\hat{b}_a \hat{b}_a^+|0\rangle = \left(1 - \hat{b}_a^+ \hat{b}_a\right)|0\rangle = |0\rangle \,.$$

In the first equality we used Eq. (1.17), in the second equality Eq. (1.20). Whence we interpret the operator $\hat{b}_a$ as the annihilation operator of the electron in the spin-orbital $|a\rangle$. Further, let us consider the action of the Hamiltonian on the state $\hat{b}_a^+ \hat{b}_b^+|0\rangle$

$$\hat{H}_e \hat{b}_a^+ \hat{b}_b^+|0\rangle = \sum_\sigma \epsilon_\sigma \hat{b}_\sigma^+ \hat{b}_\sigma \hat{b}_a^+ \hat{b}_b^+|0\rangle = \sum_\sigma \epsilon_\sigma \hat{b}_\sigma^+ \left(-\hat{b}_a^+ \hat{b}_\sigma + \delta_\sigma^a\right) \hat{b}_b^+|0\rangle =$$

$$= -\sum_\sigma \epsilon_\sigma \hat{b}_\sigma^+ \hat{b}_a^+ \left(-\hat{b}_b^+ \hat{b}_\sigma + \delta_\sigma^b\right)|0\rangle + \epsilon_a \hat{b}_a^+ \hat{b}_b^+|0\rangle = \quad (1.26)$$

$$= \left(-\epsilon_b \hat{b}_b^+ \hat{b}_a^+ + \epsilon_a \hat{b}_a^+ \hat{b}_b^+\right)|0\rangle = (\epsilon_b + \epsilon_a) \hat{b}_a^+ \hat{b}_b^+|0\rangle \,.$$

The first three equalities follow again from Eqs. (1.22) and (1.17), the fourth equality from Eq. (1.20), and the last equality from Eq. (1.19). We see that once we impose the anticommutation relation (1.17), we also have to stipulate the anticommutation relation (1.18) and vice versa. Otherwise, we could not have a non-trivial two-electron state. Equation (1.26) tells us that the state $\hat{b}_a^+ \hat{b}_b^+|0\rangle$ is the eigenstate of the Hamiltonian (1.22) with the energy $\epsilon_a + \epsilon_b$, so we see that our interpretation of the operators $\hat{b}^+$ as electron creation operators holds on. Note that, owing to the anticommutation relation (1.19), the two-electron state is antisymmetric with respect to exchange of the electrons

$$\hat{b}_a^+ \hat{b}_b^+|0\rangle = -\hat{b}_b^+ \hat{b}_a^+|0\rangle \,.$$

1.3 The One- and Two-particle Operators

If we try to put two electrons into the same spin-orbital, we get an empty state

$$\hat{b}_a^+\hat{b}_a^+|0\rangle = -\hat{b}_a^+\hat{b}_a^+|0\rangle \Rightarrow \hat{b}_a^+\hat{b}_a^+|0\rangle = 0. \quad (1.27)$$

This is the Pauli exclusion principle.

1.3 The One- and Two-particle Operators

To motivate the field form of a one-particle operator, we note that the Hamiltonian $\hat{H}_e$, Eq. (1.22), can be alternatively obtained as

$$\hat{H}_e = \int \hat{\psi}^+(r)\hat{h}\hat{\psi}(r)\,d^3r = \sum_{\sigma,\rho} \hat{b}_\sigma^+\hat{b}_\rho \int U_\sigma^+(r)\hat{h}U_\rho(r)\,d^3r = \quad (1.28)$$

$$= \sum_{\sigma,\rho} \hat{b}_\sigma^+\hat{b}_\rho \epsilon_\rho \int U_\sigma^+(r)U_\rho(r)\,d^3r = \sum_{\sigma,\rho} \hat{b}_\sigma^+\hat{b}_\rho \epsilon_\rho \delta_\rho^\sigma = \sum_\sigma \hat{b}_\sigma^+\hat{b}_\sigma \epsilon_\sigma.$$

In the second equality we used Eq. (1.23), in the third equality we used Eq. (1.10), and in the fourth equality we used Eq. (1.12).

This leads us to define the field form of a one-particle operator

$$\hat{Z} = \int \hat{\psi}^+(r)\hat{z}\hat{\psi}(r)\,d^3r = \sum_{\sigma,\rho} \hat{e}_\rho^\sigma z_\sigma^\rho, \quad (1.29)$$

where we introduced the notation for the *one-particle excitation operator*

$$\hat{e}_\rho^\sigma = \hat{b}_\sigma^+\hat{b}_\rho \quad (1.30)$$

and for one-particle matrix elements

$$z_\sigma^\rho = \langle U_\sigma|\hat{z}|U_\rho\rangle = \int U_\sigma^+(r)\hat{z}U_\rho(r)\,d^3r. \quad (1.31)$$

Henceforth, $\hat{z}$ is the one-particle operator given by Eq. (1.2).

To motivate the field form of a two-particle operator, we note the classical energy of a cloud of electric charge

$$V = \frac{1}{2}\int d^3r_1 \int d^3r_2 \frac{\rho(r_1)\rho(r_2)}{|r_1 - r_2|}.$$

If the cloud consists of N point particles, $\rho(\mathbf{r}) = \sum_{i=1}^{N} \delta(\mathbf{r} - \mathbf{r}_i)$, this energy reduces to

$$V = \frac{1}{2} \sum_{i=1}^{N} \sum_{j=1}^{N} \frac{1}{r_{ij}}.$$

If we exclude the self-interaction terms $i = j$ from the sum, this equals the two-particle operator, Eqs. (1.1) and (1.3). There is no restriction on i and j, so each pair of electrons is counted twice. The factor $1/2$ in the last expression compensates this double counting in comparison with Eqs. (1.1) and (1.3). If we substitute $\rho(\mathbf{r}) = (\psi^+\psi)(\mathbf{r})$ for the charge density and further substitute the decompositions (1.23) for the electron field and its conjugate, we obtain the field form of a two-particle operator

$$\hat{V} = \frac{1}{2} \int d^3\mathbf{r}_1 \int d^3\mathbf{r}_2 \frac{\hat{\psi}^+(\mathbf{r}_1)\hat{\psi}(\mathbf{r}_1)\hat{\psi}^+(\mathbf{r}_2)\hat{\psi}(\mathbf{r}_2)}{|\mathbf{r}_1 - \mathbf{r}_2|}$$

$$= \frac{1}{2} \sum \hat{b}_\sigma^+ \hat{b}_\mu \hat{b}_\rho^+ \hat{b}_\nu \langle \sigma|_1 \langle \rho|_2 \hat{r}_{12}^{-1} |\mu\rangle_1 |\nu\rangle_2, \qquad (1.32)$$

where

$$\langle \sigma|_1 \langle \rho|_2 \hat{r}_{12}^{-1} |\mu\rangle_1 |\nu\rangle_2 = \int d^3\mathbf{r}_1 \int d^3\mathbf{r}_2 \frac{U_\sigma^+(\mathbf{r}_1)U_\mu(\mathbf{r}_1)U_\rho^+(\mathbf{r}_2)U_\nu(\mathbf{r}_2)}{|\mathbf{r}_1 - \mathbf{r}_2|}. \qquad (1.33)$$

As pointed out above, Eq. (1.32) contains electrostatic interaction of each electron with itself. This self-interaction should be subtracted from Eq. (1.32) so that the interaction field form, Eq. (1.32), is equivalent with the interaction particle form, Eqs. (1.1) and (1.3). This is achieved by writing

$$\hat{b}_\sigma^+ \hat{b}_\mu \hat{b}_\rho^+ \hat{b}_\nu = -\hat{b}_\sigma^+ \hat{b}_\rho^+ \hat{b}_\mu \hat{b}_\nu + \delta_\mu^\rho \hat{b}_\sigma^+ \hat{b}_\nu, \qquad (1.34)$$

where Eq. (1.17) has been used. The second term on the rhs corresponds to the electron self-interaction. To see this, let us evaluate the corresponding contribution to $\hat{V}$, Eq. (1.32),

$$\hat{V}_1 = \frac{1}{2} \sum_{\sigma,\nu,\mu} \hat{b}_\sigma^+ \hat{b}_\nu \int d^3\mathbf{r}_1 \int d^3\mathbf{r}_2 \frac{U_\sigma^+(\mathbf{r}_1)U_\mu(\mathbf{r}_1)U_\mu^+(\mathbf{r}_2)U_\nu(\mathbf{r}_2)}{|\mathbf{r}_1 - \mathbf{r}_2|} = \qquad (1.35)$$

$$= \frac{4\pi}{2(2\pi)^3} \sum_{\sigma,\nu} \hat{b}_\sigma^+ \hat{b}_\nu \int d^3\mathbf{r} \int d^3\mathbf{r}_1 \int \frac{d^3\mathbf{k}}{\mathbf{k}\cdot\mathbf{k}} U_\sigma^+(\mathbf{r}_1) e^{i\mathbf{k}\cdot\mathbf{r}_1} \times$$

1.3 The One- and Two-particle Operators

$$\times \left(\sum_\mu U_\mu(r_1)U_\mu^+(r_2)\right) e^{-i\boldsymbol{k}\cdot\boldsymbol{r}_2} U_\nu(r_2) =$$

$$= \sum_\sigma \hat{b}_\sigma^+ \hat{b}_\sigma \frac{4\pi}{2(2\pi)^3} \int \frac{d^3k}{\boldsymbol{k}\cdot\boldsymbol{k}}.$$

In the second equality we used the Fourier transform of the Coulomb potential

$$\frac{1}{|r_1 - r_2|} = \frac{4\pi}{(2\pi)^3} \int \frac{d^3k}{\boldsymbol{k}\cdot\boldsymbol{k}} e^{i\boldsymbol{k}\cdot(r_1-r_2)}. \quad (1.36)$$

In the third equality we used completeness and orthonormality relations, Eqs. (1.11) and (1.12), respectively,

$$\sum_\mu U_\mu(r_1)U_\mu^+(r_2) = \delta^{(3)}(r_1 - r_2)$$

and

$$\int d^3r\, U_\sigma^+(r)U_\nu(r) = \delta_\nu^\sigma.$$

Obviously, the term (1.35) is the one-particle operator. If we add it to the Hamiltonian (1.22), all one-particle energies ϵ_σ are shifted by the same, albeit divergent, amount

$$\frac{4\pi}{2(2\pi)^3} \int \frac{d^3k}{\boldsymbol{k}\cdot\boldsymbol{k}}.$$

As it is clear from Eq. (1.36), this is the same as $2^{-1}|r_1 - r_2|^{-1}$ when $r_2 \to r_1$. We shall see that only differences of the one-particle energies appear in the experimentally measurable quantities. Thus, the electron self-interaction is of no physical significance and should be omitted, indeed.

Thus, in the following we use what is called a normal ordered form of a two-particle operator

$$:\hat{V}: = \frac{1}{2} \sum_{\sigma,\rho,\mu,\nu} \hat{e}_{\mu\nu}^{\sigma\rho} \langle\sigma|_1\langle\rho|_2 \hat{r}_{12}^{-1}|\mu\rangle_1|\nu\rangle_2 = \frac{1}{4} \sum_{\sigma,\rho,\mu,\nu} \hat{e}_{\mu\nu}^{\sigma\rho} v_{\sigma\rho}^{\mu\nu}. \quad (1.37)$$

Here we introduced the notation for the *two-particle excitation operator*

$$\hat{e}_{\mu\nu}^{\sigma\rho} = -\hat{b}_\sigma^+ \hat{b}_\rho^+ \hat{b}_\mu \hat{b}_\nu = \hat{b}_\sigma^+ \hat{b}_\rho^+ \hat{b}_\nu \hat{b}_\mu. \quad (1.38)$$

The first form comes from the rearrangement in Eq. (1.34). However, one usually uses rather the second form, where the ordering of the creation operators is the same as the ordering of the superscript indices, while the ordering of the annihilation operators is opposite to the ordering of subscript indices. Further, we introduced the antisymmetrized form of matrix elements of the Coulomb interaction

$$v_{\sigma\rho}^{\mu\nu} = \langle\sigma|_1\langle\rho|_2 \hat{r}_{12}^{-1} (|\mu\rangle_1|\nu\rangle_2 - |\nu\rangle_1|\mu\rangle_2) = (\langle\sigma|_1\langle\rho|_2 - \langle\rho|_1\langle\sigma|_2)\hat{r}_{12}^{-1}|\mu\rangle_1|\nu\rangle_2, \quad (1.39)$$

which will be used extensively in the following. The second equality is obtained by exchanging the electron coordinates 1 and 2, cf. Eq. (1.33), in the second term. The second form of Eq. (1.37) follows from the antisymmetry of the excitation operator $\hat{e}_{\mu\nu}^{\sigma\rho}$ in the upper (σ, ρ) and lower (μ, ν) indices. Thus, the field form of the Hamiltonian (1.1) that we use in the following is

$$\hat{H} = \hat{Z} + :\hat{V}:, \quad (1.40)$$

where the one- and two-particle operators are given by Eqs. (1.29) and (1.37).

Chapter 2
Hartree-Fock Approximation

When the variational wave function is searched for in the form of a single Slater determinant, the resultant variational method is called the *Hartree-Fock* method. In this chapter we introduce the notion of the Fermi vacuum and use it to derive the Hartree-Fock equations in the spin-orbital and spin-restricted forms. Further, we examine the stability of the Hartree-Fock solution.

2.1 Energy of the N-particle State

Let us start by evaluating the variational energy of a two-particle state

$$E_{\text{HF}} = \langle 0|\hat{b}_b\hat{b}_a\hat{H}\hat{b}_a^+\hat{b}_b^+|0\rangle = \langle 0|\hat{b}_b\hat{b}_a \left(\sum_{\sigma,\rho} \hat{e}_\rho^\sigma z_\sigma^\rho + \frac{1}{4} \sum_{\sigma,\rho,\mu,\nu} \hat{e}_{\mu\nu}^{\sigma\rho} v_{\sigma\rho}^{\mu\nu} \right) \hat{b}_a^+\hat{b}_b^+|0\rangle \,, \tag{2.1}$$

where we substituted for Hamiltonian from Eqs. (1.29), (1.37) and (1.40). We evaluate the average of the one- and two-particle operators as follows

$$\langle 0|\hat{b}_b\hat{b}_a\hat{e}_\rho^\sigma\hat{b}_a^+\hat{b}_b^+|0\rangle = \langle 0|\hat{b}_a\hat{e}_\rho^\sigma\hat{b}_a^+|0\rangle + \langle 0|\hat{b}_b\hat{e}_\rho^\sigma\hat{b}_b^+|0\rangle = \delta_a^\sigma \delta_\rho^a + \delta_b^\sigma \delta_\rho^b \tag{2.2}$$

and

$$\langle 0|\hat{b}_b\hat{b}_a\hat{e}_{\mu\nu}^{\sigma\rho}\hat{b}_a^+\hat{b}_b^+|0\rangle = \mathcal{A}_{\mu\nu}^{\sigma\rho} \delta_a^\sigma \delta_b^\rho \delta_\mu^a \delta_\nu^b \,, \tag{2.3}$$

where $\mathcal{A}$ designates an antisymmetrizer in the pertinent indices, for instance

$$\mathcal{A}^{\sigma\rho}\delta_a^\sigma \delta_b^\rho = \delta_a^\sigma \delta_b^\rho - \delta_a^\rho \delta_b^\sigma . \tag{2.4}$$

The first equality in Eq. (2.2) follows from the Pauli exclusion principle, Eq. (1.27): the spin-orbitals a and b have to be different from each other. This means that on the lhs of Eq. (2.2), the creation operator $\hat{b}_b^+$ has to be annihilated by the operator $\hat{b}_b$ and not by the operator $\hat{b}_a$, since a and b have to be different spin-orbitals. The rest of Eqs. (2.2) and (2.3) follows from the anticommutation relations, Eqs. (1.17), (1.18), (1.19), the vacuum definition, Eqs. (1.20) and (1.21), and the definitions of the one- and two-particle excitation operators, Eqs. (1.30) and (1.38). Further, the introduction of the antisymmetrizer not only avoids notational clutter and keeps the (anti)symmetry of the pertinent expression manifest, but also enormously simplifies the evaluation of the two-particle excitation operator between electron configurations, as we shall see later on. As it is clear from the rhs of Eq. (2.3), the antisymmetrizer in the indices σ and ρ automatically makes the expression antisymmetric in the indices a and b, too. We shall use the antisymmetrizer extensively in the following. Substituting now Eqs. (2.2) and (2.3) into Eq. (2.1) we obtain

$$E_{\text{HF}} = z_a^a + z_b^b + v_{ab}^{ab} . \tag{2.5}$$

The expressions (1.39) and (2.3) are both antisymmetric in the both pairs of the indices σ, ρ and μ, ν. This supplies twice the factor of 2, which consequently cancels the factor 2^{-2} in Eq. (2.1).

Now, the reader can convince himself that Eqs. (2.1) and (2.5) yield the same result as

$$E_{\text{HF}} = \langle \psi | \hat{H} | \psi \rangle = \frac{1}{2} \begin{vmatrix} \langle a|_1 & \langle b|_1 \\ \langle a|_2 & \langle b|_2 \end{vmatrix} \left[\hat{z}(1) + \hat{z}(2) + \hat{v}(1,2) \right] \begin{vmatrix} |a\rangle_1 & |b\rangle_1 \\ |a\rangle_2 & |b\rangle_2 \end{vmatrix} , \tag{2.6}$$

where we substituted for $|\psi\rangle$ from Eq. (1.5) and for $\hat{H}$ from Eq. (1.1).

When we evaluate the variational energy of a three-particle state, we appreciate, for the first time, the advantages of the field approach. In the field approach, we are to evaluate

$$E_{\text{HF}} = \sum_{\sigma,\rho} z_\sigma^\rho \langle 0|\hat{b}_c\hat{b}_b\hat{b}_a\hat{e}_\rho^\sigma\hat{b}_a^+\hat{b}_b^+\hat{b}_c^+|0\rangle + \frac{1}{4}\sum_{\sigma,\rho,\mu,\nu} v_{\sigma\rho}^{\mu\nu} \langle 0|\hat{b}_c\hat{b}_b\hat{b}_a\hat{e}_{\mu\nu}^{\sigma\rho}\hat{b}_a^+\hat{b}_b^+\hat{b}_c^+|0\rangle -$$

$$= \sum_{\sigma,\rho} z_\sigma^\rho \left[\langle 0|\hat{b}_a\hat{e}_\rho^\sigma\hat{b}_a^+|0\rangle + \langle 0|\hat{b}_b\hat{e}_\rho^\sigma\hat{b}_b^+|0\rangle + \langle 0|\hat{b}_c\hat{e}_\rho^\sigma\hat{b}_c^+|0\rangle \right] + \tag{2.7}$$

2.2 Fermi Vacuum

$$+ \frac{1}{4} \sum_{\sigma,\rho,\mu,\nu} v^{\mu\nu}_{\sigma\rho} \left[\langle 0| \hat{b}_b \hat{b}_a \hat{e}^{\sigma\rho}_{\mu\nu} \hat{b}^+_a \hat{b}^+_b |0\rangle + \langle 0| \hat{b}_c \hat{b}_b \hat{e}^{\sigma\rho}_{\mu\nu} \hat{b}^+_b \hat{b}^+_c |0\rangle + \right.$$

$$\left. + \langle 0| \hat{b}_c \hat{b}_a \hat{e}^{\sigma\rho}_{\mu\nu} \hat{b}^+_a \hat{b}^+_c |0\rangle \right] =$$

$$= z^a_a + z^b_b + z^c_c + v^{ab}_{ab} + v^{bc}_{bc} + v^{ac}_{ac},$$

while in the particle approach we are to evaluate

$$E_{\text{HF}} = \frac{1}{3!} \begin{vmatrix} \langle a|_1 & \langle b|_1 & \langle c|_1 \\ \langle a|_2 & \langle b|_2 & \langle c|_2 \\ \langle a|_3 & \langle b|_3 & \langle c|_3 \end{vmatrix} \left[\hat{z}(1) + \hat{z}(2) + \hat{z}(3) + \hat{v}(1,2) + \hat{v}(2,3) + \hat{v}(1,3) \right] \times$$

$$\times \begin{vmatrix} |a\rangle_1 & |b\rangle_1 & |c\rangle_1 \\ |a\rangle_2 & |b\rangle_2 & |c\rangle_2 \\ |a\rangle_3 & |b\rangle_3 & |c\rangle_3 \end{vmatrix}.$$

In the second equality in Eq. (2.7) we again used the Pauli exclusion principle, Eq. (1.27), that is, the spin-orbitals a, b and c have to differ from each other. In the third equality we used Eqs. (2.2) and (2.3). Although the field and particle approaches yield the same result, in the particle approach we are to consider $3!^3 = 216$ terms during the calculation.

From Eqs. (2.5) and (2.7) we are to guess that the variational energy of the N-particle state $\hat{b}^+_N \ldots \hat{b}^+_1 |0\rangle$ is

$$E_{\text{HF}} = \sum_{a=1}^{N} z^a_a + \sum_{a=1}^{N} \sum_{b=a+1}^{N} v^{ab}_{ab}. \tag{2.8}$$

This guess is correct, but we are to introduce the so-called hole-particle formalism, which makes the guess unnecessary.

2.2 Fermi Vacuum

From the atomic shell model, we know that the ground state of N-electrons is dominated by a single configuration. For instance, the ground state of helium is dominated by the configuration where the two electrons are in the hydrogen-like 1s state with spins up and down. The ground state of lithium is dominated by the helium configuration plus the third electron in the hydrogen-like 2s state and so on. From the success of the atomic shell model, we guess that also the ground state of a molecule is dominated by a single configuration. We will consider this single configuration as a new vacuum state, the so-called *Fermi vacuum*,

$$|F\rangle = \hat{b}_N^+ \ldots \hat{b}_1^+ |0\rangle. \tag{2.9}$$

If we act on the Fermi vacuum by the creation operator of an occupied spin-orbital, we get by virtue of the Pauli exclusion principle, Eq. (1.27), the empty state

$$\hat{b}_a^+ |F\rangle = 0. \tag{2.10}$$

On the other hand, if we act on the Fermi vacuum by the annihilation operator of an unoccupied or virtual spin-orbital, we get the empty state as well, as follows from Eq. (1.20)

$$\hat{b}_r |F\rangle = 0. \tag{2.11}$$

Further, the Fermi vacuum state is normalized, cf. Eqs. (1.17) and (1.27),

$$\langle F|F\rangle = 1. \tag{2.12}$$

Now, we can forget about the original definition of the Fermi vacuum, Eq. (2.9), and use the last three equations as our working definition of the Fermi vacuum. In an analogy with quantum electrodynamics, we shall call the creation of an electron in a virtual orbital as the creation of a particle and the annihilation of an electron in an occupied orbital as the creation of a hole.

As it will be clear in due time, it is advantageous to put the Hamiltonian (1.40) into the normally ordered form with respect to the new Fermi vacuum. The normal ordering is usually defined in such a way that the creation operators are on the left of the annihilation operators. That means that no contraction of the pertinent indices is allowed in a normally ordered expression. A contraction of the pertinent indices means that, according to Eq. (1.17), the pertinent operators are contracted to a number—the Kronecker delta. Let us illustrate it on a one-particle operator; we have

$$\hat{e}_\rho^\sigma = \hat{b}_\sigma^+ \hat{b}_\rho =: \hat{e}_\rho^\sigma : + h(\rho)\delta_\rho^\sigma, \tag{2.13}$$

where the function $h(\mu)$ equals 1 if μ designates an occupied spin-orbital and 0 if μ designates a virtual spin-orbital. If the indices σ and ρ are to be contracted, their ordering must be "wrong" with respect to the Fermi vacuum. However, for this to be the case, they both have to designate an occupied spin-orbital. The operator $\hat{b}_\rho$ then creates a hole and $\hat{b}_\sigma^+$ annihilates it. Likewise, if we have a two-particle operator, then, by virtue of Eq. (2.13), we have

$$\hat{e}_{\mu\nu}^{\sigma\rho} = \hat{b}_\sigma^+ \hat{b}_\rho^+ \hat{b}_\nu \hat{b}_\mu = \hat{b}_\sigma^+ \left(: \hat{e}_\nu^\rho : + h(\nu)\delta_\nu^\rho\right) \hat{b}_\mu. \tag{2.14}$$

2.2 Fermi Vacuum

In the first term on the rhs, the index ν cannot be contracted with the index ρ, but it can still be contracted with the index σ

$$\hat{b}_\sigma^+ : \hat{e}_\nu^\rho : \hat{b}_\mu = -h(\nu)\delta_\nu^\sigma \hat{e}_\mu^\rho + : \hat{b}_\sigma^+ \hat{b}_\rho^+ \hat{b}_\nu : \hat{b}_\mu \,.$$

In the second term on the rhs, the index ν cannot be contracted anymore, but the index μ can

$$: \hat{b}_\sigma^+ \hat{b}_\rho^+ \hat{b}_\nu : \hat{b}_\mu =: \hat{e}_{\mu\nu}^{\sigma\rho} : + h(\mu) \mathcal{A}^{\sigma\rho} \delta_\mu^\sigma : \hat{e}_\nu^\rho : .$$

Substituting the last two equations into Eq. (2.14) we obtain

$$\hat{e}_{\mu\nu}^{\sigma\rho} =: \hat{e}_{\mu\nu}^{\sigma\rho} : + h(\nu)\mathcal{A}^{\rho\sigma} \delta_\nu^\rho \hat{e}_\mu^\sigma + h(\mu)\mathcal{A}^{\rho\sigma}\delta_\mu^\sigma : \hat{e}_\nu^\rho : .$$

Using once again Eq. (2.13) in the second term on the rhs, we finally get

$$\hat{e}_{\mu\nu}^{\sigma\rho} = h(\nu)h(\mu)\mathcal{A}^{\rho\sigma}\delta_\nu^\rho \delta_\mu^\sigma + h(\nu)\mathcal{A}^{\rho\sigma}\delta_\nu^\rho : \hat{e}_\mu^\sigma : + h(\mu)\mathcal{A}^{\rho\sigma}\delta_\mu^\sigma : \hat{e}_\nu^\rho + : \hat{e}_{\mu\nu}^{\sigma\rho} : . \tag{2.15}$$

Substituting Eqs. (2.13) and (2.15) into Eqs. (1.29), (1.37) and (1.40), we can express the N-particle Hamiltonian as a sum of a number—vacuum expectation value—and a Hamiltonian in a normally ordered form with respect to the Fermi vacuum

$$\hat{H} = <\hat{H}> + : \hat{H} : . \tag{2.16}$$

The first term on the rhs of the last equation is

$$<\hat{H}> = \langle F|\hat{H}|F\rangle = \sum z_\rho^\sigma h(\rho)\delta_\rho^\sigma + \frac{1}{4}\sum v_{\sigma\rho}^{\mu\nu} \mathcal{A}^{\sigma\rho} h(\nu)h(\mu)\delta_\nu^\rho \delta_\mu^\sigma = \tag{2.17}$$

$$= \sum_\rho h(\rho)\left[z_\rho^\rho + \frac{1}{2}\sum_\sigma h(\sigma)v_{\sigma\rho}^{\sigma\rho}\right] = E_{\text{HF}}.$$

In the third equality we multiply the terms $v_{\sigma\rho}^{\mu\nu}$ and $\mathcal{A}^{\sigma\rho}h(\nu)h(\mu)\delta_\nu^\rho \delta_\mu^\sigma$. They both are antisymmetric in the indices σ and ρ. As the reader can easily verify, this is the same as leaving out the explicit antisymmetrization in the latter expression and multiplying the whole expression by the factor of two. Note that this reasoning was already used in Eq. (2.5) and we shall use it extensively in the following. The last equality follows from the comparison of Eqs. (2.8) and (2.17).

The second term in (2.16) reads

$$:\hat{H} := \sum :\hat{e}^\sigma_\rho: z^\rho_\sigma + \frac{1}{4}\sum \left(\mathcal{A}^{\sigma\rho}h(\nu)\delta^\rho_\nu :\hat{e}^\sigma_\mu: + \right.$$
$$\left. + \mathcal{A}^{\sigma\rho}h(\mu)\delta^\sigma_\mu :\hat{e}^\rho_\nu: + :\hat{e}^{\sigma\rho}_{\mu\nu}:\right) v^{\mu\nu}_{\sigma\rho} = \quad (2.18)$$

$$= \sum :\hat{e}^\sigma_\rho: f^\rho_\sigma + \frac{1}{4}\sum :\hat{e}^{\sigma\rho}_{\mu\nu}: v^{\mu\nu}_{\sigma\rho},$$

where we introduced the matrix elements of the Fock operator

$$f^\rho_\sigma = z^\rho_\sigma + \sum_\nu h(\nu) v^{\rho\nu}_{\sigma\nu}. \quad (2.19)$$

Recall that $h(\mu)$ is non-zero for occupied spin-orbitals only. Therefore, in practice, the sum on the rhs of the last equation runs through all occupied spin-orbitals only.

2.3 Hartree-Fock Equations

The Hartree-Fock spin-orbitals are obtained from the requirement that the Hartree-Fock energy functional, Eqs. (2.8) or (2.17), is minimal. As is well-known, the necessary, but not sufficient, condition for a minimum of an energy functional is that the variation with respect to a spin-orbital vanishes

$$\frac{\delta E_{\text{HF}}}{\delta \langle a|} = \frac{\delta \epsilon_a}{\delta \langle a|} = 0, \quad (2.20)$$

where

$$\epsilon_a = \frac{\langle a|\hat{f}|a\rangle}{\langle a|a\rangle}. \quad (2.21)$$

The first equality follows from the fact that the energy functional, Eq. (2.8), depends on the particular spin-orbital a only through the combination $z^a_a + \sum_b v^{ab}_{ab}$. However, given the definition of the Fock matrix, Eq. (2.19), this is precisely $\langle a|\hat{f}|a\rangle$. Also, we have been assuming the spin-orbitals are normalized, but clearly we can leave out this restriction if we divide the average of the Fock operator by the square of the spin-orbital norm $\langle a|a\rangle$. Assuming the Hermitian conjugate spin-orbitals are independent on spin-orbitals, we have

$$\frac{\delta \frac{\langle a|\hat{f}|a\rangle}{\langle a|a\rangle}}{\delta \langle a|} = \frac{\hat{f}|a\rangle}{\langle a|a\rangle} - \frac{\langle a|\hat{f}|a\rangle}{\langle a|a\rangle^2}|a\rangle = 0 \Rightarrow \hat{f}|a\rangle = \epsilon_a|a\rangle. \quad (2.22)$$

2.3 Hartree-Fock Equations

Thus, the necessary condition for the minimization of the Hartree-Fock energy functional, Eq. (2.8), results in the diagonalization of the Fock operator, Eqs. (2.19) and (2.22).

To find the relation between the one-particle energies ϵ_a and the Hartree-Fock energy E_{HF}, we note that Eqs. (2.19) and (2.21) yield

$$\sum_{a=1}^{N} \epsilon_a = \sum_{a=1}^{N} \frac{z_a^a + \sum_{b=1}^{N} v_{ab}^{ab}}{\langle a | a \rangle}.$$

A comparison of the last equation and Eq. (2.8) yields the sought relation

$$E_{\text{HF}} = \frac{1}{2} \sum_{a=1}^{N} \left[\epsilon_a + \frac{\langle a | \hat{z} | a \rangle}{\langle a | a \rangle} \right]. \tag{2.23}$$

On the practical level, the spin-orbitals a are expanded in fixed, complete and entirely discrete one-particle basis,

$$|a\rangle = \sum_{j=1}^{B} c_i^{(a)} |j\rangle. \tag{2.24}$$

Multiplying Eq. (2.22) successively by B vectors $\langle i|$, Eq. (2.22) is transformed into the matrix eigenvalue problem

$$\sum_{j=1}^{B} f_i^j c_j^{(a)} = \epsilon_a \sum_{j=1}^{B} S_i^j c_j^{(a)}, \tag{2.25}$$

where S_{ij} and f_{ij} designate overlap and Fock operator matrix elements

$$S_i^j = \langle i | j \rangle \tag{2.26}$$

and

$$f_{ij} = z_i^j + \sum_{b=1}^{N} v_{ib}^{jb} = z_i^j + \sum_{b=1}^{N} \frac{\sum_{k=1}^{B} \sum_{l=1}^{B} c_k^{(b)} c_l^{(b)} v_{ik}^{jl}}{\sum_{k=1}^{B} \sum_{l=1}^{B} c_k^{(b)} c_l^{(b)} S_k^l}, \tag{2.27}$$

respectively. In the Hartree-Fock approximation, an electron is moving in a field of nuclei and an averaged field of the other electrons. As it is clear from Eqs. (2.25) and (2.27), the determination of the field due to the other electrons is a part of the problem. To determine the field of the other electrons, we have to determine the coefficients c, cf. Eq. (2.27). However, these coefficients form the eigenvalue vectors of the Fock matrix that we are to calculate, cf. Eq. (2.25). The trick here is to use an iterative self-consistent approach. We start with a reasonable initial guess of the coefficients c, evaluate the Fock matrix, Eq. (2.27), diagonalize the Fock matrix, Eq. (2.25), and obtain B eigenvalues and eigenvectors. We identify the N eigenvectors corresponding to the lowest N eigenvalues with the new guess of the c coefficients, evaluate the new Fock matrix, diagonalize it again, and so on. We continue this process until we reach self-consistency. In practice, this means until the N lowest eigenvalues ϵ_a do not significantly change.

The quality of any Hartree-Fock and post-Hartree-Fock calculation strongly depends on the form and the number of the basis spin-orbitals, Eq. (2.24). However, we will not discuss this important issue here and refer reader to the book [1] instead.

2.4 Spin-restricted Form of Hartree-Fock Equations

If the Hamiltonian is non-relativistic, see, e.g., Eqs. (1.1)–(1.3), then it is spin-independent. Consequently, the Hartree-Fock equations (2.19) and (2.22) can be written in terms of orbitals rather than spin-orbitals. Consider, for instance, the beryllium atom. There are two electrons with opposite spins in the $1s$ orbital and two electrons with opposite spins in the $2s$ orbital

$$|1\rangle = |1s\rangle|+\rangle, |2\rangle = |1s\rangle|-\rangle, |3\rangle = |2s\rangle|+\rangle, |4\rangle = |2s\rangle|-\rangle. \qquad (2.28)$$

Further, consider Eq. (2.22) for, for example, the spin-orbital $|1\rangle$. Equation (2.19) then yields

$$\hat{f}|1\rangle = \hat{z}|1\rangle + \langle 2|_2 \hat{r}_{12}^{-1} (|1\rangle_1|2\rangle_2 - |2\rangle_1|1\rangle_2) +$$
$$+ \langle 3|_2 \hat{r}_{12}^{-1} (|1\rangle_1|3\rangle_2 - |3\rangle_1|1\rangle_2) + \langle 4|_2 \hat{r}_{12}^{-1} (|1\rangle_1|4\rangle_2 - |4\rangle_1|1\rangle_2).$$

Multiplying this equation by the spin state $\langle +|$ from the left, using the spin-independence of $\hat{z}$ and taking into account the form of the spin-orbitals given by penultimate equation, we obtain

$$\hat{f}|1s\rangle = \hat{z}|1s\rangle + \langle 1s|_2 \hat{r}_{12}^{-1} |1s\rangle_1 |1s\rangle_2 +$$
$$+ \langle 2s|_2 \hat{r}_{12}^{-1} (|1s\rangle_1|2s\rangle_2 - |2s\rangle_1|1s\rangle_2) + \langle 2s|_2 \hat{r}_{12}^{-1} |1s\rangle_1 |2s\rangle_2 =$$
$$= \hat{z}|1s\rangle + \langle 1s|_2 \hat{r}_{12}^{-1} |1s\rangle_1 |1s\rangle_2 + \langle 2s|_2 \hat{r}_{12}^{-1} (2|1s\rangle_1|2s\rangle_2 - |2s\rangle_1|1s\rangle_2).$$

The action of the Fock operator on the $2s$ orbital is obtained from the last equation by reversing the roles of $1s$ and $2s$.

2.5 Stability Conditions

Thus, for a general closed-shell system comprising $2N$ electrons in N orbitals, the spin-restricted Hartree-Fock equations read

$$\hat{f}|A\rangle = \epsilon_A |A\rangle ,$$

where

$$\hat{f}|A\rangle = \hat{z}_1|A\rangle_1 + \sum_{B=1}^{N} \langle B|_2 \hat{r}_{12}^{-1} \left(2|A\rangle_1 |B\rangle_2 - |B\rangle_1 |A\rangle_2\right) . \quad (2.29)$$

2.5 Stability Conditions

As noted above, the requirement that the first variation of the energy functional vanishes is a necessary, but not a sufficient condition for the found solution to be indeed an energy minimum. To get also the necessary condition for at a least local minimum of an energy functional, we have to examine the second variation.[1]

Consider the variational energy

$$E_{\text{var}} - E_{\text{HF}} = \frac{\langle \psi | : \hat{H} : | \psi \rangle}{\langle \psi | \psi \rangle} , \quad (2.30)$$

for a general monoexcited state written in the form of an exponential of a so-called cluster operator, cf. the next chapter,[2]

$$|\psi\rangle = \exp\{\hat{T}_1\}|F\rangle , \quad \hat{T}_1 = \sum_{a=1}^{N} \sum_{r=N+1}^{B} t_a^r \hat{e}_a^r , \quad (2.31)$$

where a and r designate occupied and virtual spin-orbitals, respectively. The coefficients t_a^r are linear combinations of the decomposition of the cluster operator $\hat{T}_1$ in terms of creation and annihilation operators. The stability conditions we search for will be written in terms of these coefficients, see Eqs. (2.34) and (2.37) below. A normally ordered Hamiltonian takes the form, cf. Eq. (2.18),

$$: \hat{H} := \sum : \hat{e}_\sigma^\sigma : \epsilon_\sigma + \frac{1}{4} \sum : \hat{e}_{\mu\nu}^{\sigma\rho} : v_{\sigma\rho}^{\mu\nu} . \quad (2.32)$$

---

[1] This and the following Sections lie somewhat out of the main development and can be omitted upon the first reading. The stability conditions for the Hartree-Fock solution were for the first time discussed by David Thouless (1934–2019) [15]. The following spin adaption is due to Čížek and Paldus [16].

[2] See also, for instance, [5], for a proof that general monoexcited state can be indeed written in this manner.

Here we substituted, cf. Eq. (2.22),

$$f_\sigma^\rho = \delta_\sigma^\rho \epsilon_\sigma \, .$$

Substituting Eq. (2.31) into Eq. (2.30) and expanding the resulting expression in the powers of $\hat{T}_1$, we get up to the second order in $\hat{T}_1$

$$E_{\text{var}} - E_{\text{HF}} \simeq \left[\langle F | : \hat{H} : | F \rangle + \langle F | \left(: \hat{H} : \hat{T}_1 + \hat{T}_1^+ : \hat{H} : \right) | F \rangle + \right.$$

$$\left. + \frac{1}{2} \langle F | \left(: \hat{H} : \hat{T}_1^2 + 2\hat{T}_1^+ : \hat{H} : \hat{T}_1 + (\hat{T}_1^+)^2 : \hat{H} : \right) | F \rangle \right] \times$$

$$\times \left[\langle F | F \rangle + \langle F | (\hat{T}_1 + \hat{T}_1^+) | F \rangle + \frac{1}{2} \langle F | \left(\hat{T}_1^2 + 2\hat{T}_1^+ \hat{T}_1 + (\hat{T}_1^+)^2 \right) | F \rangle \right]^{-1} ,$$

The absolute and linear terms in $\hat{T}_1$ in the numerator vanish due to the way normally ordered Hamiltonian is constructed, see Eqs. (2.16), (2.17) and (2.32). For instance,

$$\langle F | \left(: \hat{H} : \hat{T}_1 \right) | F \rangle = \sum_{a=1}^{N} \sum_{r=N+1}^{B} t_a^r \left(\sum_\mu \epsilon_\mu \langle F | : \hat{e}_\mu^\mu : \hat{e}_a^r | F \rangle + \right.$$

$$\left. + \frac{1}{4} \sum_{\sigma,\rho,\mu,\nu} v_{\sigma\rho}^{\mu\nu} \langle F | : \hat{e}_{\mu\nu}^{\sigma\rho} : \hat{e}_a^r | F \rangle \right) = 0 \, .$$

Here, we used the fact that in a normally ordered operator indices cannot be contracted between themselves. We used also the equalities

$$\langle F | : \hat{e}_{\mu\nu}^{\sigma\rho} : \hat{e}_a^r | F \rangle = 0$$

and

$$\langle F | : \hat{e}_\mu^\mu : \hat{e}_a^r | F \rangle = \delta_a^r = 0 \, ,$$

which hold due to the fact that a and r label states from different subspaces of the one-particle state; a and r label occupied and virtual spin-orbitals, respectively, $a \in (1, N)$ and $r \in (N + 1, B)$. Thus, with accuracy up to the second order in $\hat{T}_1$, we have, using Eq. (2.12),

$$E_{\text{var}} - E_{\text{HF}} \simeq + \frac{1}{2} \langle F | \left(: \hat{H} : \hat{T}_1^2 + 2\hat{T}_1^+ : \hat{H} : \hat{T}_1 + (\hat{T}_1^+)^2 : \hat{H} : \right) | F \rangle \, . \quad (2.33)$$

Substituting for the normally ordered Hamiltonian from Eqs. (2.32) and for the cluster operator (2.31), we have for the Hamiltonian matrix elements between the

2.5 Stability Conditions

monoexcited states

$$\langle F|\hat{T}_1^+ : \hat{H} : \hat{T}_1|F\rangle = \sum t_a^r(t_b^s)^* \left(\epsilon_\mu \langle F|\hat{e}_s^b : \hat{e}_\mu^\mu : \hat{e}_a^r|F\rangle + \right.$$
$$\left. + \frac{1}{4} v_{\sigma\rho}^{\mu\nu} \langle F|\hat{e}_s^b : \hat{e}_{\mu\nu}^{\sigma\rho} : \hat{e}_a^r|F\rangle \right) =$$
$$= \sum t_a^r(t_b^s)^* \left[\delta_a^b \delta_s^r (\epsilon_r - \epsilon_a) + v_{sa}^{br}\right],$$

since

$$\langle F|\hat{e}_s^b : \hat{e}_\mu^\mu : \hat{e}_a^r|F\rangle = \delta_a^b \delta_s^\mu \delta_\mu^r - \delta_s^r \delta_a^\mu \delta_\mu^b$$

and

$$\langle F|\hat{e}_s^b : \hat{e}_{\mu\nu}^{\sigma\rho} : \hat{e}_a^r|F\rangle = -\mathcal{A}_{\mu\nu}^{\sigma\rho} \delta_v^r \delta_s^\sigma \delta_b^\mu \delta_a^\rho .$$

Further, since the Hamilton operator is Hermitian, we have

$$\langle F|(\hat{T}_1^+)^2 : \hat{H} : |F\rangle = \left(\langle F| : \hat{H} : \hat{T}_1^2|F\rangle\right)^*$$

and the Hamiltonian matrix elements between the Fermi vacuum and biexcited states read

$$\langle F| : \hat{H} : \hat{T}_1^2|F\rangle = \sum t_a^r t_b^s \frac{1}{4} v_{\sigma\rho}^{\mu\nu} \langle F| : \hat{e}_{\mu\nu}^{\sigma\rho} : \hat{e}_b^s \hat{e}_a^r|F\rangle = \sum t_a^r t_b^s v_{ab}^{rs},$$

as

$$\langle F| : \hat{e}_{\mu\nu}^{\sigma\rho} : \hat{e}_b^s \hat{e}_a^r|F\rangle = -\mathcal{A}_{\mu\nu}^{\sigma\rho} \delta_\mu^s \delta_a^\sigma \delta_v^r \delta_b^\rho .$$

Thus, the expansion of the variational energy in the vicinity of the Hartree-Fock solution, Eq. (2.33), can be written in the form

$$E_{\text{var}} - E_{\text{HF}} \simeq \frac{1}{2} \sum \left(t_a^r \ (t_a^r)^*\right) \begin{pmatrix} A_{as}^{rb} & B_{ab}^{rs} \\ (B_{ab}^{rs})^* & (A_{as}^{rb})^* \end{pmatrix} \begin{pmatrix} (t_b^s)^* \\ t_b^s \end{pmatrix} . \tag{2.34}$$

The matrices A and B are given as

$$A_{as}^{rb} = \delta_a^b \delta_s^r (\epsilon_r - \epsilon_a) + v_{sa}^{br} \tag{2.35}$$

and

$$B_{ab}^{rs} = v_{ab}^{rs} . \tag{2.36}$$

We see that the Hartree-Fock solution is a local minimum in the space of one-particle states if all eigenvalues Λ of the stability matrix

$$\begin{pmatrix} \hat{A} & \hat{B} \\ \hat{B}^* & \hat{A}^* \end{pmatrix} \begin{pmatrix} t^* \\ t \end{pmatrix} = \Lambda \begin{pmatrix} t^* \\ t \end{pmatrix} \tag{2.37}$$

are positive as then the rhs of Eq. (2.34) is positive.

2.6 Spin Adapted Stability Matrix

There are two simplifications of these stability conditions. Firstly, as the non-relativistic molecular Hamiltonian is real, cf. Eqs. (1.1)–(1.3), the stationary solution of the Schrödinger equation can be chosen to be real. Consequently, all the matrix elements displayed above are real and the eigenvalue problem for stability matrix reduces to

$$\hat{\Lambda} t = \Lambda t, \quad \Lambda^{rb}_{as} = A^{rb}_{as} + B^{rs}_{ab} = \delta^b_a \delta^r_s (\epsilon_r - \epsilon_a) + 2 v^{br}_{sa}. \tag{2.38}$$

Here we used that for real spin-orbitals $v^{rs}_{ab} = v^{br}_{sa}$, cf. Eq. (1.39). Secondly, for a spin-independent Hamiltonian, the monoexcitations can be chosen to be the eigenstates of the square and of the third component of an electron field spin operator

$$\hat{S} = S^\nu_\mu \hat{e}^\mu_\nu, \tag{2.39}$$

where S^ν_μ are the matrix elements of a one-electron spin operator; explicitly[3]

$$S^\nu_\mu = \langle \mu | \frac{\sigma}{2} | \nu \rangle, \quad \sigma = \left(\begin{pmatrix} 0 & 1 \\ 1 & 0 \end{pmatrix}, \begin{pmatrix} 0 & -i \\ i & 0 \end{pmatrix}, \begin{pmatrix} 1 & 0 \\ 0 & -1 \end{pmatrix} \right). \tag{2.40}$$

Putting now the spin operator in the normally ordered form with respect to the Fermi vacuum, we obtain, cf. Eq. (2.13),

$$\hat{S} = S^\nu_\mu \left(: \hat{e}^\mu_\nu : + h(\nu) \delta^\mu_\nu \right) = S^\nu_\mu : \hat{e}^\mu_\nu :, \tag{2.41}$$

since we assume that the Fermi vacuum is a closed-shell state where all total spin projections vanish. As is well-known, see, e.g., [14], it is advantageous to introduce

---

[3] Let us recall that we use the atomic units in which $\hbar = 1$.

2.6 Spin Adapted Stability Matrix

non-Hermitian combinations of the spin operators

$$\hat{S}_{\pm} = \hat{S}_x \pm i\hat{S}_y . \qquad (2.42)$$

The square of the spin operator can then be written

$$\hat{S}^2 = \hat{S}_x^2 + \hat{S}_y^2 + \hat{S}_z^2 = \frac{1}{2}\left(\hat{S}_+\hat{S}_- + \hat{S}_-\hat{S}_+\right) + \hat{S}_z^2 .$$

For a spin-independent Hamiltonian, the Hartree-Fock spin-orbitals can be written as a direct product of orbital and spin states, cf. Eq. (2.28),

$$|\mu\rangle = |M\rangle|2m\rangle , \; m = \pm\frac{1}{2} . \qquad (2.43)$$

Substituting this into Eqs. (2.39), (2.40) and (2.42), we find that the components of the electron field spin operator read

$$: \hat{S}_z := \frac{1}{2}\sum_M \left(: \hat{e}_{M,+}^{M,+} : - : \hat{e}_{M,-}^{M,-} :\right)$$

$$: \hat{S}_+ := \sum_M : \hat{e}_{M,-}^{M,+} : , \quad : \hat{S}_- := \sum_M : \hat{e}_{M,+}^{M,-} : ,$$

since, for instance,

$$(S_z)_\mu^\nu = \langle M|\langle 2m|\frac{\sigma_z}{2}|N\rangle|2n\rangle = \delta_M^N \langle 2m|\frac{\sigma_z}{2}|2n\rangle = \delta_M^N \delta_m^n \frac{1}{2}\left(\delta_{2m,1} - \delta_{2m,-1}\right) .$$

For the action of these operators on particle and hole states, we find

$$: \hat{S}_z : \hat{b}_{R,\pm}^+|F\rangle = \pm\frac{1}{2}\hat{b}_{R,\pm}^+|F\rangle ,$$

$$: \hat{S}_z : \hat{b}_{A,\pm}|F\rangle = \mp\frac{1}{2}\hat{b}_{A,\pm}|F\rangle ,$$

$$: \hat{S}_+ : \hat{b}_{R,+}^+|F\rangle = 0 , \quad : \hat{S}_+ : \hat{b}_{R,-}^+|F\rangle = \hat{b}_{R,+}^+|F\rangle ,$$

$$: \hat{S}_- : \hat{b}_{R,+}^+|F\rangle = \hat{b}_{R,-}^+|F\rangle , \quad : \hat{S}_- : \hat{b}_{R,-}^+|F\rangle = 0 ,$$

$$: \hat{S}_+ : \hat{b}_{A,+}|F\rangle = -\hat{b}_{A,-}|F\rangle , \quad : \hat{S}_+ : \hat{b}_{A,-}|F\rangle = 0$$

and

$$: \hat{S}_- : \hat{b}_{A,+}|F\rangle = 0 , \quad : \hat{S}_- : \hat{b}_{A,-}|F\rangle = -\hat{b}_{A,+}|F\rangle .$$

We see that the action of the spin operators on the holes is exactly opposite to that on the particles. For the action of the spin operators on the monoexcitations, we get from the above equations

$$:\hat{S}_z : \hat{b}^+_{R,+}\hat{b}_{A,+}|F\rangle = 0,$$

$$:\hat{S}_+ : \hat{b}^+_{R,+}\hat{b}_{A,+}|F\rangle = -\hat{b}^+_{R,+}\hat{b}_{A,-}|F\rangle,$$

$$:\hat{S}_- : \hat{b}^+_{R,+}\hat{b}_{A,-}|F\rangle = \hat{b}^+_{R,-}\hat{b}_{A,-}|F\rangle - \hat{b}^+_{R,+}\hat{b}_{A,+}|F\rangle,$$

$$:\hat{S}_- : \hat{b}^+_{R,+}\hat{b}_{A,+}|F\rangle = \hat{b}^+_{R,-}\hat{b}_{A,+}|F\rangle$$

and

$$:\hat{S}_+ : \hat{b}^+_{R,-}\hat{b}_{A,+}|F\rangle = \hat{b}^+_{R,+}\hat{b}_{A,+}|F\rangle - \hat{b}^+_{R,-}\hat{b}_{A,-}|F\rangle.$$

Thus, for the action of the square of the spin operator, we get from the foregoing equations

$$\hat{S}^2 \hat{b}^+_{R,+}\hat{b}_{A,+}|F\rangle = \hat{b}^+_{R,+}\hat{b}_{A,+}|F\rangle - \hat{b}^+_{R,-}\hat{b}_{A,-}|F\rangle$$

and by up-down symmetry

$$\hat{S}^2 \hat{b}^+_{R,-}\hat{b}_{A,-}|F\rangle = \hat{b}^+_{R,-}\hat{b}_{A,-}|F\rangle - \hat{b}^+_{R,+}\hat{b}_{A,+}|F\rangle.$$

Whence the eigenstates of the square of the spin operator are the singlet states

$$\hat{S}^2 \frac{1}{\sqrt{2}}(\hat{b}^+_{R,+}\hat{b}_{A,+} + \hat{b}^+_{R,-}\hat{b}_{A,-})|F\rangle = 0$$

and triplet states

$$\hat{S}^2 \frac{1}{\sqrt{2}}(\hat{b}^+_{R,+}\hat{b}_{A,+} - \hat{b}^+_{R,-}\hat{b}_{A,-})|F\rangle = 2\frac{1}{\sqrt{2}}(\hat{b}^+_{R,+}\hat{b}_{A,+} - \hat{b}^+_{R,-}\hat{b}_{A,-})|F\rangle.$$

The stability matrix, Eq. (2.38), does not mix these two groups of states together. Inserting separation (2.43) into the elements of the stability matrix, Eq. (2.38), one obtains

$$\Lambda^{(R,\pm)(B,\pm)}_{(A,\pm)(S,\pm)} = \delta^B_A \delta^R_S (\epsilon_R - \epsilon_A) + 2v^{BR}_{SA}$$

and

$$\Lambda^{(R,\pm)(B,\mp)}_{(A,\pm)(S,\mp)} = 2\langle S|_1 \langle A|_2 \hat{r}^{-1}_{12}|B\rangle_1|R\rangle_2.$$

2.6 Spin Adapted Stability Matrix

We form the singlet and triplet stability matrices

$$\Lambda_{AS}^{RB}(0) = \Lambda_{(A,\pm)(S,\pm)}^{(R,\pm)(B,\pm)} + \Lambda_{(A,\pm)(S,\mp)}^{(R,\pm)(B,\mp)} =$$

$$= \delta_A^B \delta_S^R (\epsilon_R - \epsilon_A) + \langle S|_1 \langle A|_2 \hat{r}_{12}^{-1} (4|B\rangle_1 |R\rangle_2 - 2|R\rangle_1 |B\rangle_2)$$

and

$$\Lambda_{AS}^{RB}(1) = \Lambda_{(A,\pm)(S,\pm)}^{(R,\pm)(B,\pm)} - \Lambda_{(A,\pm)(S,\mp)}^{(R,\pm)(B,\mp)} =$$

$$= \delta_A^B \delta_S^R (\epsilon_R - \epsilon_A) - 2 \langle S|_1 \langle A|_2 \hat{r}_{12}^{-1} |R\rangle_1 |B\rangle_2 ,$$

respectively. If one of the eigenvalues of the singlet or triplet matrix is negative, one has a singlet or a triplet instability, respectively.

A triplet instability indicates that there is a Fermi vacuum with a broken spin symmetry, i.e., with a non-zero vacuum expectation value of $\hat{S}^2$, below the spin symmetric Fermi vacuum with $\hat{S}^2|F\rangle = 0$. This means that a configuration where the electrons with opposite spins occupy different orbitals is energetically more favorable than the symmetric configuration, where the electrons with opposite spins occupy the same orbitals. This happens, for instance, in the hydrogen molecule for large internuclear separations, cf. the discussion in [1]. For a large internuclear separation, the state when the first electron occupies $1s$ state at the nucleus A with spin up and the second electron occupies $1s$ state at the nucleus B with spin down (or vice versa) has lower energy than the state when both electrons occupy the same molecular orbital 1σ. So the asymmetric Fermi vacuum $|F_A\rangle = \hat{b}_{1s_A,+}^+ \hat{b}_{1s_B,-}^+ |0\rangle$ has lower energy than the symmetric Fermi vacuum $|F_S\rangle = \hat{b}_{1\sigma,+}^+ \hat{b}_{1\sigma,-}^+ |0\rangle$. For the action of $\hat{S}^2$ on these Fermi vacua we have

$$\hat{S}^2 |F_S\rangle = 0$$

and

$$\hat{S}^2 \hat{b}_{1s_A,+}^+ \hat{b}_{1s_B,-}^+ |0\rangle = \hat{b}_{1s_A,+}^+ \hat{b}_{1s_B,-}^+ |0\rangle + \hat{b}_{1s_A,-}^+ \hat{b}_{1s_B,+}^+ |0\rangle .$$

Clearly, one can introduce a symmetry restored Fermi vacuum

$$|F_R\rangle = \hat{b}_{1s_A,+}^+ \hat{b}_{1s_B,-}^+ |0\rangle - \hat{b}_{1s_A,-}^+ \hat{b}_{1s_B,+}^+ |0\rangle , \quad \hat{S}^2 |F_R\rangle = 0 .$$

This way, however, one goes beyond the Hartree-Fock model and no longer writes the N-electron state in the form of a single Slater determinant.

A singlet instability means that either there is a spatial-symmetry-broken Fermi vacuum bellow the Fermi vacuum respecting a spatial symmetry of the fixed nuclear framework or there is no Hartree-Fock solution at all for the fixed nuclear framework, see, e.g., [17]. While for systems with a finite number of electrons, the spin or spatial symmetry breaking are artefacts of the independent particle model and do not appear in the exact solution, the purely singlet instability preserving all the symmetries of Hamiltonian is a strong indication that the fixed nuclear framework with the fixed number of electrons does not support any bound state at all.

Note that these stability conditions are rarely used. The reason is that usually if there is a closely lying symmetry-broken solution or there is no solution of the Hartree-Fock equations for the fixed nuclear framework, the iterative self-consistent method of solution described below Eq. (2.27) does not converge. Nonetheless, the foregoing considerations are of some practical significance. It has long been recognized[4] that the spin instability is always present in an open-shell system. Consider, for instance, the lithium atom with the configuration

$$|1\rangle = |1s+\rangle|+\rangle \, , \, |2\rangle = |1s-\rangle|-\rangle \, , \, |3\rangle = |2s\rangle|+\rangle \, .$$

Equation (2.19) yields

$$\hat{f}|1s+\rangle = \hat{z}|1s+\rangle + \langle 1s-|_2\hat{r}_{12}^{-1}|1s+\rangle_1|1s-\rangle_2 + \langle 2s+|_2\hat{r}_{12}^{-1}(|1s+\rangle_1|2s+\rangle_2 - |2s+\rangle_1|1s+\rangle_2)$$

and

$$\hat{f}|1s-\rangle = \hat{z}|1s-\rangle + \langle 1s+|_2\hat{r}_{12}^{-1}|1s-\rangle_1|1s+\rangle_2 + \langle 2s+|_2\hat{r}_{12}^{-1}|1s-\rangle_1|2s+\rangle_2 \, .$$

Clearly, due to the extra electron in the $2s$ state with spin up, the Fock operators for the electrons in the $1s$ state with spin up and spin down are different. The Pauli principle forbids electrons with the same spin to approach each other, but not with opposite spins. One can still set $|1s+\rangle = |1s-\rangle$, but such a solution lies energetically higher than the one with different $|1s+\rangle$ and $|1s-\rangle$ orbitals.

If one encounters an instability in the Hartree-Fock solution, one is left with basically four options. First, one works further with the symmetric, yet unstable Hartree-Fock solution. However, one can then expect convergence problems with the iterative solution of the coupled cluster method to be described in the following Chap. 4, cf. [9]. Second, one works further with a stable, but symmetry-broken Hartree-Fock solution. However, one then works with an approximate solution that does not share the symmetry of the exact solution. Third, one works with a symmetry-restored Hartree-Fock solution. However, one then works with a reference state comprising more than one Slater determinant. Fourth, in case of an open-shell neutral system, one solves the Hartree-Fock equations for a closed-shell cation. The extra open-shell electron is put into the first excited orbital obtained

---

[4] Perhaps the first to state this explicitly was Per-Olov Löwdin (1916–2000), see [18].

2.6 Spin Adapted Stability Matrix

from the closed-shell solution. This is called the frozen core approximation. Each of these options has its drawbacks, but unfortunately that is all we have.

In conclusion, instabilities of the Hartree-Fock solutions considerably complicate further considerations. In the following we shall assume that the Hartree-Fock solution does exist and is stable. See for instance [4] for a review of the methods how to proceed further if this is not the case.

Chapter 3
Coupled Cluster Method

In this chapter we come to the exposition of the coupled cluster method. We first describe the method of configuration interaction and discuss its main shortcoming: the size-extensivity problem. Then we show how the coupled cluster exponential Ansatz solves this problem in a systematic way. Finally, we derive the spin-orbital form of equations for coupled cluster amplitudes.

3.1 Configuration Interaction

The Hartree-Fock method yields B optimized one-particle spin-orbitals, which is now our model space. Of these B spin-orbitals, the energetically lowest N spin-orbitals are occupied. Thus, the Hartree-Fock method provides us with the basis in the N-particle Hilbert space

$$\mathbf{1} = |0\rangle\langle 0| + |2\rangle\langle 2| + |4\rangle\langle 4| + |6\rangle\langle 6| + |8\rangle\langle 8| + \ldots, \qquad (3.1)$$

where we have made the symbolic assignments

$$|0\rangle \leftrightarrow |F\rangle, \quad \langle 0| \leftrightarrow \langle F|, \quad |2\rangle \leftrightarrow \hat{e}_a^r |F\rangle, \quad \langle 2| \leftrightarrow \langle F|\hat{e}_a^r, \qquad (3.2)$$

$$|4\rangle \leftrightarrow \frac{1}{2!^2}\hat{e}_{ab}^{rs}|F\rangle, \quad \langle 4| \leftrightarrow \langle F|\hat{e}_{rs}^{ab}, \quad |6\rangle \leftrightarrow \frac{1}{3!^2}\hat{e}_{abc}^{rst}|F\rangle, \quad \langle 6| \leftrightarrow \langle F|\hat{e}_{rst}^{abc},$$

$$|8\rangle \leftrightarrow \frac{1}{4!^2}\hat{e}_{abcd}^{rstu}|F\rangle, \quad \langle 8| \leftrightarrow \langle F|\hat{e}_{rstu}^{abcd},$$

and so on. Here, the letters from the beginning of the alphabet ($a, b, c,\ldots$) refer to the occupied spin-orbitals $a \in (1, N)$ and the letters from the middle of the alphabet ($r, s, t,\ldots$) refer to the virtual spin-orbitals, $r \in (N + 1, B)$. The two-particle states $|2\rangle$ are called *monoexcitations* since one electron is excited to create a hole in the occupied states and a particle in the virtual states, the four-particle states $|4\rangle$ are called *biexcitations* since two electrons are excited to create two holes in the occupied states and two particles in the virtual states, and so on. The factorial factors in the assignment of the ket-vectors are due to the normalization, see, e.g., Eq. (3.11) below.

If we multiply Eq. (3.1) by the exact state, $|\psi\rangle$, we find that the exact state is given as sum over all possible excitation from Hartree-Fock configuration

$$|\psi\rangle = c_0|F\rangle + \sum_{a=1}^{N}\sum_{r=N+1}^{B} c_a^r \hat{e}_a^r |F\rangle + \frac{1}{2!^2}\sum_{a=1}^{N}\sum_{b=1}^{N}\sum_{r=N+1}^{B}\sum_{s=N+1}^{B} c_{ab}^{rs}\hat{e}_{ab}^{rs}|F\rangle + \qquad (3.3)$$

$$+ \frac{1}{3!^2}\sum_{a=1}^{N}\sum_{b=1}^{N}\sum_{c=1}^{N}\sum_{r=N+1}^{B}\sum_{s=N+1}^{B}\sum_{t=N+1}^{B} c_{abc}^{rst}\hat{e}_{abc}^{rst}|F\rangle +$$

$$+ \frac{1}{4!^2}\sum_{a=1}^{N}\sum_{b=1}^{N}\sum_{c=1}^{N}\sum_{d=1}^{N}\sum_{r=N+1}^{B}\sum_{s=N+1}^{B}\sum_{t=N+1}^{B}\sum_{u=N+1}^{B} c_{abcd}^{rstu}\hat{e}_{abcd}^{rstu}|F\rangle + \dots$$

This can be written in a shorthand symbolic form

$$|\psi\rangle = c_0|0\rangle + c_2|2\rangle + c_4|4\rangle + c_6|6\rangle + c_8|8\rangle + \dots, \qquad (3.4)$$

where obviously

$$c_0 = \langle 0|\psi\rangle, \; c_2 = \langle 2|\psi\rangle, \; c_4 = \langle 4|\psi\rangle \qquad (3.5)$$

and so on.

For further considerations it is advantageous to substitute Eqs. (2.16) and (2.17) into the Schrödinger equation

$$\hat{H}|\psi\rangle = E|\psi\rangle, \qquad (3.6)$$

and rewrite it into slightly more convenient form

$$:\hat{H}:|\psi\rangle = E_{\text{corr}}|\psi\rangle, \qquad (3.7)$$

where

$$E_{\text{corr}} = E - E_{\text{HF}} \qquad (3.8)$$

3.1 Configuration Interaction

is called the *correlation energy* and the normal ordered Hamiltonian has the form (2.32).

Substituting now the expansion (3.4) into Eq. (3.7) and multiplying the equation successively from the left by $\langle 0|$, $\langle 2|$, $\langle 4|$ and so on, we obtain

$$\begin{pmatrix} 0 & 0 & H_{04} & 0 & 0 \\ 0 & H_{22} & H_{24} & H_{26} & 0 \\ H_{40} & H_{42} & H_{44} & H_{46} & H_{48} \\ 0 & H_{62} & H_{64} & H_{66} & H_{68} \\ 0 & 0 & H_{84} & H_{86} & H_{88} \end{pmatrix} \begin{pmatrix} c_0 \\ c_2 \\ c_4 \\ c_6 \\ c_8 \end{pmatrix} = E_{\text{corr}} \begin{pmatrix} c_0 \\ c_2 \\ c_4 \\ c_6 \\ c_8 \end{pmatrix}. \quad (3.9)$$

Firstly, we tacitly assumed that the higher than eight-particle terms on the rhs of Eq. (3.1) can be neglected. Secondly, we used the so-called *Brillouin theorem*. The Hamiltonian matrix elements between the Fermi vacuum and the monoexcited states vanish

$$H_{02} \leftrightarrow \langle F| : \hat{H} : \hat{e}_a^r |F\rangle = \langle F| : \left(\sum : \hat{e}_\sigma^\sigma : \epsilon_\sigma + \frac{1}{4} \sum : \hat{e}_{\mu\nu}^{\sigma\rho} : v_{\sigma\rho}^{\mu\nu} \right) \hat{e}_a^r |F\rangle = $$
$$= \sum_\sigma \epsilon_\sigma \delta_\sigma^r \delta_a^\sigma = \epsilon_a \delta_a^r = 0, \quad (3.10)$$

since indices a and r belong to disjoint subspaces of the model space and can never equal each other. Thirdly, we used the so-called *Slater rules*: the Hamiltonian matrix elements between $2m$ and $2n$ particle configurations vanish, $H_{2n2m} = 0$, whenever $|m - n| > 2$. For instance

$$H_{06} \leftrightarrow \frac{1}{3!^2} \langle F| : \hat{H} : \hat{e}_{abc}^{rst} |F\rangle =$$
$$= \frac{1}{3!^2} \langle F| \left(\sum : \hat{e}_\sigma^\sigma : \epsilon_\sigma + \frac{1}{4} \sum : \hat{e}_{\mu\nu}^{\sigma\rho} : v_{\sigma\rho}^{\mu\nu} \right) \hat{e}_{abc}^{rst} |F\rangle = 0,$$

since obviously there are not enough free indices to be contracted with the excitation operator indices. Lastly, on the rhs of Eq. (3.9) we used

$$\langle 2|2\rangle c_2 \leftrightarrow \frac{1}{2!^2} \sum_{abrs} \langle F|\hat{e}_{tu}^{cd}\hat{e}_{ab}^{rs}|F\rangle c_{ab}^{rs} = \frac{1}{2!^2} \sum_{abrs} \Delta_{ab}^{cd} \Delta_{tu}^{rs} c_{ab}^{rs} = c_{cd}^{tu} \leftrightarrow c_2 \quad (3.11)$$

and so on. Here, we introduced the antisymmetrized Kronecker delta

$$\Delta_{ab}^{cd} = \mathcal{A}^{cd} \delta_a^c \delta_b^d = \delta_a^c \delta_b^d - \delta_a^d \delta_b^c \quad (3.12)$$

and used the antisymmetry of the c coefficients, cf. Eq. (3.3),

$$c_{ab}^{rs} = -c_{ba}^{rs} = -c_{ab}^{sr} = c_{ba}^{sr}. \tag{3.13}$$

Now, we can find the correlation energy by solving the eigenvalue problem, Eq. (3.9). This is called the method of *configuration interaction*. However, we find that for a large enough number of basis functions B and a large enough number of particles N, the combinatorics overwhelm us. The number of n-particle states grows rapidly with increasing n. Nonetheless, as it is clear from Eq. (3.9), the Hartree-Fock state, i.e., the Fermi vacuum, interacts directly with the four-particle states only. In fact, for small enough systems, we obtain good results by considering the Fermi vacuum and four-particle states only,

$$\begin{pmatrix} 0 & H_{04} \\ H_{40} & H_{44} \end{pmatrix} \begin{pmatrix} c_0 \\ c_4 \end{pmatrix} = E_{\text{corr}} \begin{pmatrix} c_0 \\ c_4 \end{pmatrix}. \tag{3.14}$$

However, for large enough B and N, we encounter the problem called *size extensivity*.

3.2 Problem of Size Extensivity

The problem of size extensivity can be most simply stated as follows.[1] Let a molecule consists of two non-interacting identical atoms A and B. The Hamiltonian $:\hat{\mathcal{H}}:$ of the molecule then can be written in the form of a sum of the atomic Hamiltonians

$$:\hat{\mathcal{H}}:=:\hat{H}_A: \otimes 1_B + 1_A \otimes :\hat{H}_B:=:\hat{H}_A: + :\hat{H}_B:, \tag{3.15}$$

where in the second equality we switched to the physicists sloppy notation. The atomic Hamiltonians are supposed to be of the same form, but act on different states: $\hat{H}_A$ on the states of the atom A, $\hat{H}_B$ on the states of the atom B. We take the atomic Hamiltonian as simple as possible; its Hilbert space consists of the Fermi vacuum and one four-particle state. Thus, the configuration interaction takes the form of Eq. (3.14), where $H_{04} = K/2$, $H_{40} = 2K$ and $H_{44} = 2\Delta$, where K and Δ are real numbers,

$$\begin{pmatrix} 0 & K/2 \\ 2K & 2\Delta \end{pmatrix} \begin{pmatrix} c_0 \\ c_4 \end{pmatrix} = E_{\text{corr}} \begin{pmatrix} c_0 \\ c_4 \end{pmatrix}, \tag{3.16}$$

$$E_{\text{corr}} = \Delta \pm \sqrt{\Delta^2 + K^2}. \tag{3.17}$$

---

[1] The following argument is a slightly modified version of the argument given in [1].

3.2 Problem of Size Extensivity

When finding the eigenvalues of the molecular Hamiltonian $:\hat{\mathcal{H}}:$,

$$:\hat{\mathcal{H}}:|\psi\rangle = \mathcal{E}|\psi\rangle, \tag{3.18}$$

we expect to find, as according to our assumptions the atoms A and B are identical and non-interacting, the size extensive result

$$\mathcal{E} = 2E_{\text{corr}}. \tag{3.19}$$

As we shall see, in an approximate calculation, this is not necessarily the case.

We expand the solution of the Schrödinger equation, Eq. (3.18), in the basis of our model

$$|\psi\rangle = \sum_{i=1}^{4} c_i |i\rangle,$$

where

$$|1\rangle = |0\rangle_A |0\rangle_B, \quad |2\rangle = |0\rangle_A |4\rangle_B, \quad |3\rangle = |4\rangle_A |0\rangle_B, \quad |4\rangle = |4\rangle_A |4\rangle_B$$

and multiply Eq. (3.18) successively by $\langle j|$, $j = 1, \ldots, 4$ from the left. We obtain

$$\begin{pmatrix} 0 & K/2 & K/2 & 0 \\ 2K & 2\Delta & 0 & K/2 \\ 2K & 0 & 2\Delta & K/2 \\ 0 & 2K & 2K & 4\Delta \end{pmatrix} \begin{pmatrix} c_1 \\ c_2 \\ c_3 \\ c_4 \end{pmatrix} = \mathcal{E} \begin{pmatrix} c_1 \\ c_2 \\ c_3 \\ c_4 \end{pmatrix}, \tag{3.20}$$

$$\mathcal{E} = \left\{ 2\left(\Delta \pm \sqrt{\Delta^2 + K^2}\right), 2\Delta, 2\Delta \right\}, \tag{3.21}$$

where we substituted, for instance,

$$\langle 1|\hat{\mathcal{H}}|2\rangle = \langle 0|_A \langle 0|_B (\hat{H}_A + \hat{H}_B)|0\rangle_A |4\rangle_B = H_{04} = K/2,$$

$$\langle 2|\hat{\mathcal{H}}|2\rangle = \langle 0|_A \langle 4|_B (\hat{H}_A + \hat{H}_B)|0\rangle_A |4\rangle_B = H_{00} + H_{44} = 0 + 2\Delta,$$

and so on. Clearly, the size extensivity requirement, Eq. (3.19), is fulfilled for the lowest eigenvalues of atomic and molecular Hamiltonians, Eqs. (3.17) and (3.21). However, had we restricted calculation to biexcited configurations, i.e., we would exclude the tetraexcited configuration $|4\rangle$, the size extensivity, Eq. (3.19), would not hold.

3.3 Coupled Cluster Equations in Matrix Form

The considerations in previous Section raise the question if an approximate calculation can be done in such a way that the size extensivity requirement is always fulfilled. This brings us to the coupled cluster method. The basic trick how to ensure that the size extensivity requirement is fulfilled is to search for a solution of the Schrödinger equation (3.7) in an exponential form

$$|\psi\rangle = \exp\{\hat{T}\}|F\rangle, \quad (3.22)$$

where $\hat{T}$ is a non-Hermitian operator whose precise form will be specified in due time. Substituting Eq. (3.22) into Eq. (3.7) and multiplying the resulting equation by $\exp\{-\hat{T}\}$ from the left, we obtain

$$\hat{\bar{H}}|F\rangle = E_{\text{corr}}|F\rangle, \quad \hat{\bar{H}} = \exp\{-\hat{T}\} : \hat{H} : \exp\{\hat{T}\}. \quad (3.23)$$

Returning back to our model example discussed in the previous section, we stipulate the operator $\hat{T}$ has the only non-vanishing matrix element between the biexcited state and the Fermi vacuum

$$\hat{T} = \begin{pmatrix} T_{00} & T_{04} \\ T_{40} & T_{44} \end{pmatrix} = \begin{pmatrix} 0 & 0 \\ t & 0 \end{pmatrix}. \quad (3.24)$$

The coupled cluster Ansatz, Eq. (3.22), means that we search for solution of Eq. (3.16) in the form

$$\begin{pmatrix} c_0 \\ c_4 \end{pmatrix} = \exp\left\{\begin{pmatrix} 0 & 0 \\ t & 0 \end{pmatrix}\right\} \begin{pmatrix} 1 \\ 0 \end{pmatrix} = \begin{pmatrix} 1 & 0 \\ t & 1 \end{pmatrix} \begin{pmatrix} 1 \\ 0 \end{pmatrix}.$$

Substituting the last equation into Eq. (3.16) and multiplying the resulting equation by the operator $\exp\{-\hat{T}\} = \begin{pmatrix} 1 & 0 \\ -t & 1 \end{pmatrix}$ from the left, we obtain

$$\begin{pmatrix} 1 & 0 \\ -t & 1 \end{pmatrix} \begin{pmatrix} 0 & K/2 \\ 2K & 2\Delta \end{pmatrix} \begin{pmatrix} 1 & 0 \\ t & 1 \end{pmatrix} \begin{pmatrix} 1 \\ 0 \end{pmatrix} = E_{\text{corr}} \begin{pmatrix} 1 \\ 0 \end{pmatrix}.$$

Multiplying the matrices we obtain

$$\begin{pmatrix} Kt/2 \\ -Kt^2/2 + 2\Delta t + 2K \end{pmatrix} = \begin{pmatrix} E_{\text{corr}} \\ 0 \end{pmatrix}.$$

3.3 Coupled Cluster Equations in Matrix Form

The latter of these equations is the quadratic equation for the cluster amplitude t. Its solution is

$$t_{1,2} = 2\frac{\Delta \pm \sqrt{\Delta^2 + K^2}}{K}. \tag{3.25}$$

By substituting this solution into the former of these equations, $E_{\text{corr}} = Kt/2$, we find the solution (3.16) for the correlation energy.

For the molecular Hamiltonian, Eqs. (3.15) and (3.20), the operator $\hat{T}$ takes the form

$$\hat{T} = \hat{T}_A + \hat{T}_B = \begin{pmatrix} 0 & 0 & 0 & 0 \\ t & 0 & 0 & 0 \\ t & 0 & 0 & 0 \\ 0 & t & t & 0 \end{pmatrix}, \tag{3.26}$$

where we substituted, for instance,

$$\langle 2|\hat{T}|1\rangle = \langle 0|_A \langle 4|_B (\hat{T}_A + \hat{T}_B)|0\rangle_A |0\rangle_B = t,$$

where in the last equality we used Eq. (3.24). Searching now for the solution of Eq. (3.20) in the form

$$\begin{pmatrix} c_1 \\ c_2 \\ c_3 \\ c_4 \end{pmatrix} = \exp\{\hat{T}\} \begin{pmatrix} 1 \\ 0 \\ 0 \\ 0 \end{pmatrix} = \begin{pmatrix} 1 & 0 & 0 & 0 \\ t & 1 & 0 & 0 \\ t & 0 & 1 & 0 \\ t^2 & t & t & 1 \end{pmatrix} \begin{pmatrix} 1 \\ 0 \\ 0 \\ 0 \end{pmatrix},$$

and multiplying the resulting equation by $\exp\{-\hat{T}\}$ from the left, we obtain

$$\begin{pmatrix} 1 & 0 & 0 & 0 \\ -t & 1 & 0 & 0 \\ -t & 0 & 1 & 0 \\ t^2 & -t & -t & 1 \end{pmatrix} \begin{pmatrix} 0 & K/2 & K/2 & 0 \\ 2K & 2\Delta & 0 & K/2 \\ 2K & 0 & 2\Delta & K/2 \\ 0 & 2K & 2K & 4\Delta \end{pmatrix} \begin{pmatrix} 1 & 0 & 0 & 0 \\ t & 1 & 0 & 0 \\ t & 0 & 1 & 0 \\ t^2 & t & t & 1 \end{pmatrix} \begin{pmatrix} 1 \\ 0 \\ 0 \\ 0 \end{pmatrix} = \mathcal{E} \begin{pmatrix} 1 \\ 0 \\ 0 \\ 0 \end{pmatrix},$$

where for t we take the lower of the two solutions, Eq. (3.25). After multiplying the matrices we finally get

$$\begin{pmatrix} 2\left(\Delta - \sqrt{\Delta^2 + K^2}\right) & K/2 & K/2 & 0 \\ 0 & 2\Delta & 0 & K/2 \\ 0 & 0 & 2\Delta & K/2 \\ 0 & 0 & 0 & 2\left(\Delta + \sqrt{\Delta^2 + K^2}\right) \end{pmatrix} \begin{pmatrix} 1 \\ 0 \\ 0 \\ 0 \end{pmatrix} = \mathcal{E} \begin{pmatrix} 1 \\ 0 \\ 0 \\ 0 \end{pmatrix}.$$

One can thus see that the coupled cluster Ansatz, Eqs. (3.22) and (3.23), brings the molecular Hamiltonian into a non-Hermitian triangle form. In this form, the size extensivity requirement, Eq. (3.19), is fulfilled automatically, i.e., irrespective of how we truncate the molecular Hamiltonian matrix.

In general, the non-Hermitian operator $\hat{T}$ in Eq. (3.22) is written as the sum of one-body, two-body, three-body, etc., clusters

$$\hat{T} = \hat{T}_1 + \hat{T}_2 + \hat{T}_3 + \ldots, \tag{3.27}$$

where[2]

$$\hat{T}_1 = \sum_{a,r} t_a^r \hat{e}_a^r, \tag{3.28}$$

$$\hat{T}_2 = \sum_{a,b,r,s} \frac{1}{2!^2} t_{ab}^{rs} \hat{e}_{ab}^{rs}, \tag{3.29}$$

and so on. As we have noted above the most important is the contribution of the biexcited states. Therefore, we restrict ourselves to the two-body clusters only,

$$\hat{T} \simeq \hat{T}_2. \tag{3.30}$$

It is not a problem to account for the one-body clusters as well, i.e., to take $\hat{T} \simeq \hat{T}_1 + \hat{T}_2$, as the number of monoexcited configurations is small. However when the Hartree-Fock solution is stable, the contribution of the one-body clusters is known to be small. Nonetheless, three-body clusters are usually taken into account in a perturbative manner. This is only to estimate their contribution, since their number is very high. Higher-body clusters are usually neglected completely. It is reasonable to expect their contribution to be negligible, although it is very hard to make general statements.[3] The approximation (3.30) is referred to as CCD—coupled clusters with doubles. In this chapter we proceed further within the approximation (3.30). Inclusion of the one- and three-body clusters is described in the following chapter.

---

[2] A historical remark is perhaps in order here. As emphasized by Rodney Bartlett in his perspective paper, the last reference in [3], the exponential form of the exact multi-electron state, Eq. (3.22), was suggested by a number of physicists, most notably by John Hubbard (1931–1980), Fritz Coester (1921–2020) and Hermann Kümmel (1922–2012). However, it was only Jiří Čížek who already in his 1965 doctoral thesis written in Czech, *Příspěvek ke studiu korelačních efektů u atomů a molekul*, and then in his 1966 paper [3], took the next crucial step to expand the cluster operator in terms of the field operators, Eqs. (3.28) and (3.29), and to get in the approximation (3.30) the explicit equations for the cluster amplitudes, Eq. (3.40) below.

[3] See discussion in [9].

3.3 Coupled Cluster Equations in Matrix Form

To proceed further within the approximation (3.30), we assume that the Fermi vacuum and four- and eight-particle states form an approximately complete basis set

$$\hat{1} \simeq |0\rangle\langle 0| + |4\rangle\langle 4| + |8\rangle\langle 8|. \tag{3.31}$$

Multiplying the last equation by the state vector $|\psi\rangle$ from the right we get

$$|\psi\rangle = c_0|0\rangle + c_4|4\rangle + c_8|8\rangle,$$

where clearly

$$c_0 = \langle 0|\psi\rangle, \quad c_4 = \langle 4|\psi\rangle, \quad c_8 = \langle 8|\psi\rangle.$$

The Schrödinger equation, Eq. (3.7), in this basis reads

$$\begin{pmatrix} 0 & H_{04} & 0 \\ H_{40} & H_{44} & H_{48} \\ 0 & H_{84} & H_{88} \end{pmatrix} \begin{pmatrix} c_0 \\ c_4 \\ c_8 \end{pmatrix} = E_{\text{corr}} \begin{pmatrix} c_0 \\ c_4 \\ c_8 \end{pmatrix}. \tag{3.32}$$

The matrix elements of the cluster operator, Eqs. (3.29) and (3.30), in this basis read

$$\hat{T} = \begin{pmatrix} 0 & 0 & 0 \\ T_{40} & 0 & 0 \\ 0 & T_{84} & 0 \end{pmatrix}, \quad \hat{T}^2 = \begin{pmatrix} 0 & 0 & 0 \\ 0 & 0 & 0 \\ T_{84}T_{40} & 0 & 0 \end{pmatrix}. \tag{3.33}$$

The coupled-cluster exponential Ansatz, Eq. (3.22), then takes the form

$$\begin{pmatrix} c_0 \\ c_4 \\ c_8 \end{pmatrix} = \begin{pmatrix} 1 & 0 & 0 \\ T_{40} & 1 & 0 \\ \frac{T_{84}T_{40}}{2} & T_{84} & 1 \end{pmatrix} \begin{pmatrix} 1 \\ 0 \\ 0 \end{pmatrix}.$$

Inserting now the last equation into Eq. (3.32) and multiplying the resulting equation by the inverse exponential of the cluster operator, we obtain

$$\begin{pmatrix} 1 & 0 \\ -T_{40} & 1 \end{pmatrix} \begin{pmatrix} 0 & H_{04} & 0 \\ H_{40} & H_{44} & H_{48} \end{pmatrix} \begin{pmatrix} 1 \\ T_{40} \\ \frac{T_{84}T_{40}}{2} \end{pmatrix} = E_{\text{corr}} \begin{pmatrix} 1 \\ 0 \end{pmatrix}.$$

The last equation represents a system of two equations. The first equation yields the relation between the correlation energy and the cluster amplitudes,

$$H_{04} T_{40} = E_{\text{corr}} \tag{3.34}$$

and the second equation yields an equation for the cluster amplitudes

$$H_{40} + H_{44}T_{40} - T_{40}H_{04}T_{40} + \frac{1}{2}H_{48}T_{84}T_{40} = 0. \quad (3.35)$$

When solving the eigenvalue problem like (3.32), we first determine the eigenvalues E_{corr} and then eigenvectors, i.e., the c coefficients. However, the strategy is reversed in the coupled cluster approach; from Eq. (3.35) we determine the cluster amplitudes and then from Eq. (3.34) the correlation energy.

3.4 Coupled Cluster Equations in Spin-orbital Form

Using now the approximate decomposition of unity, Eq. (3.31), and the correspondence (3.2), the last two equations read

$$\langle F | : \hat{H} : \hat{T}_2 | F \rangle = E_{\text{corr}} \quad (3.36)$$

and

$$0 = \langle F | \hat{e}_{rs}^{ab} \left[: \hat{H} : + : \hat{H} : \hat{T}_2 - \hat{T}_2 : \hat{H} : \hat{T}_2 + \frac{1}{2} : \hat{H} : \hat{T}_2^2 \right] | F \rangle. \quad (3.37)$$

Henceforth we switch to the Einstein summation convention: if one index appears more than once, it is automatically summed over, i.e., there is no summation sign. For instance, in this convention, the expressions (2.32) and (3.29) for the Hamilton and cluster operators, read

$$: \hat{H} := \hat{e}_\sigma^\sigma : \epsilon_\sigma + \frac{1}{4} : \hat{e}_{\mu\nu}^{\sigma\rho} : v_{\sigma\rho}^{\mu\nu} \quad (3.38)$$

and

$$\hat{T}_2 = \frac{1}{2!^2} t_{ab}^{rs} \hat{e}_{ab}^{rs}, \quad (3.39)$$

respectively. Using these expressions for the Hamilton and cluster operators, the anticommutation relations (1.17)–(1.19), and the definition of the Fermi vacuum state, Eqs. (2.10)–(2.11), we obtain from Eq. (3.37) the spin-orbital form of the coupled equations for the cluster amplitudes (they will be derived in detail below)

$$e_{rs}^{ab} = 0, \quad e_{rs}^{ab} = v_{rs}^{ab} + \frac{1}{2^2} t_{cd}^{uv} \mathcal{L}_{rs,cd}^{ab,uv} + \frac{1}{2^5} t_{cd}^{uv} t_{ef}^{xy} \mathcal{Q}_{rs,cd,ef}^{ab,uv,xy}, \quad (3.40)$$

3.4 Coupled Cluster Equations in Spin-orbital Form

where

$$\mathcal{L}^{ab,uv}_{rs,cd} = \langle F | \hat{e}^{ab}_{rs} : \hat{H} : \hat{e}^{uv}_{cd} | F \rangle = (\epsilon_r + \epsilon_s - \epsilon_a - \epsilon_b) \Delta^{uv}_{rs} \Delta^{ab}_{cd} + \quad (3.41)$$

$$+ v^{ab}_{cd} \Delta^{uv}_{rs} + v^{uv}_{rs} \Delta^{ab}_{cd} + \mathcal{A}^{ab}_{rs} \mathcal{A}^{uv}_{cd} v^{au}_{rc} \delta^v_s \delta^b_d$$

and

$$\mathcal{Q}^{ab,uv,xy}_{rs,cd,ef} = \langle F | \hat{e}^{ab}_{rs} \left[-2\hat{e}^{uv}_{cd} : \hat{H} : + : \hat{H} : \hat{e}^{uv}_{cd} \right] \hat{e}^{xy}_{ef} | F \rangle = \quad (3.42)$$

$$= \mathcal{A}^{uv}_{cd} \mathcal{A}^{xy}_{ef} \Delta^{vx}_{rs} \Delta^{ab}_{de} v^{uy}_{cf} + 2\mathcal{A}^{uv} \mathcal{A}^{xy} \Delta^{ux}_{rs} \Delta^{ab}_{cd} v^{yv}_{ef} - 2\mathcal{A}_{cd} \mathcal{A}_{ef} \Delta^{xy}_{rs} \Delta^{ab}_{ed} v^{uv}_{cf} +$$

$$+ 2\Delta^{xy}_{rs} \Delta^{ab}_{cd} v^{uv}_{ef} .$$

In a similar manner, we get from Eq. (3.36) an explicit relation between the correlation energy and the cluster amplitudes

$$E_{\text{corr}} = \frac{1}{4} v^{ab}_{rs} t^{rs}_{ab} . \quad (3.43)$$

Notice that all we need in the end are the Fermi vacuum and the four-particle excitations from it only. Nonetheless, the inclusion of the eight-particle excitations at the intermediate stage of the calculation is precisely what distinguishes the coupled cluster approach from the configuration interaction method. Had we restricted ourselves to the Fermi vacuum and four-particle excitations from the very beginning, we would be missing the last term on the lhs of Eq. (3.35). That is, Eq. (3.35) would read

$$H_{40} + H_{44} T_{40} - T_{40} H_{04} T_{40} = 0 . \quad (3.44)$$

The system of Eqs. (3.34) and (3.44) then yields the same correlation energies as the configuration interaction with the Fermi vacuum and four-particle states, Eq. (3.14).

Let us derive Eqs. (3.40)–(3.43) in some detail. Substituting Eqs. (3.38) and (3.39) into Eq. (3.36) we get

$$E_{\text{corr}} = \frac{1}{4^2} v^{\kappa\lambda}_{\mu\nu} t^{rs}_{ab} \langle F | \hat{e}^{\mu\nu}_{\kappa\lambda} \hat{e}^{rs}_{ab} | F \rangle = \frac{1}{4} v^{ab}_{rs} t^{rs}_{ab} .$$

In the last equality we used first, recall Eq. (3.12),

$$\langle F | \hat{e}^{\mu\nu}_{\kappa\lambda} \hat{e}^{rs}_{ab} | F \rangle = \Delta^{rs}_{\kappa\lambda} \Delta^{\mu\nu}_{ab} ,$$

and second the antisymmetry of the potential matrix elements, cf. Eq. (1.39),

$$v^{\mu\nu}_{\sigma\rho} = -v^{\mu\nu}_{\rho\sigma} = -v^{\nu\mu}_{\sigma\rho} = v^{\nu\mu}_{\rho\sigma}. \qquad (3.45)$$

Likewise we have, substituting Eq. (3.38) into the first term on the rhs of Eq. (3.37),

$$\langle F|\hat{e}^{ab}_{rs} : \hat{H} : |F\rangle = \frac{1}{4} v^{\kappa\lambda}_{\mu\nu} \langle F|\hat{e}^{ab}_{rs} \hat{e}^{\mu\nu}_{\kappa\lambda}|F\rangle = v^{ab}_{rs}.$$

Further, we substitute for the normally ordered Hamiltonian from Eq. (3.38) into the second term on the rhs of Eq. (3.37); we obtain the expression (3.41)

$$\langle F|\hat{e}^{ab}_{rs} : \hat{H} : \hat{e}^{uv}_{cd}|F\rangle = \epsilon_\mu \langle F|\hat{e}^{ab}_{rs} : \hat{e}^{\mu}_{\mu} : \hat{e}^{uv}_{cd}|F\rangle + \frac{1}{4} v^{\kappa\lambda}_{\mu\nu} \langle F|\hat{e}^{ab}_{rs} : \hat{e}^{\mu\nu}_{\kappa\lambda} : \hat{e}^{uv}_{cd}|F\rangle. \qquad (3.46)$$

To evaluate the one-particle term we note

$$\langle F|\hat{e}^{ab}_{rs} : \hat{e}^{\mu}_{\mu} : \hat{e}^{uv}_{cd}|F\rangle = \Delta^{uv}_{rs} \langle F|\hat{b}^+_a \hat{b}^+_b : \hat{b}^+_\mu \hat{b}_\mu : \hat{b}_d \hat{b}_c|F\rangle +$$
$$+ \Delta^{ab}_{cd} \langle F|\hat{b}_s \hat{b}_r : \hat{b}^+_\mu \hat{b}_\mu : \hat{b}^+_u \hat{b}^+_v|F\rangle.$$

The remaining vacuum averages are

$$\langle F|\hat{b}^+_a \hat{b}^+_b : \hat{b}^+_\mu \hat{b}_\mu : \hat{b}_d \hat{b}_c|F\rangle = \mathcal{A}^{ab}_{cd} \delta^\mu_c \delta^d_\mu \delta^a_d$$

and

$$\langle F|\hat{b}_s \hat{b}_r : \hat{b}^+_\mu \hat{b}_\mu : \hat{b}^+_u \hat{b}^+_v|F\rangle = \mathcal{A}^{uv}_{rs} \delta^\mu_r \delta^u_\mu \delta^v_s$$

Substituting the last three equations into the first term on the rhs of Eq. (3.46), we obtain

$$\epsilon_\mu \langle F|\hat{e}^{ab}_{rs} : \hat{e}^{\mu}_{\mu} : \hat{e}^{uv}_{cd}|F\rangle = (\epsilon_r + \epsilon_s - \epsilon_a - \epsilon_b) \Delta^{uv}_{rs} \Delta^{ab}_{cd}. \qquad (3.47)$$

To evaluate the two-particle term we note

$$\langle F|\hat{e}^{ab}_{rs} : \hat{e}^{\mu\nu}_{\kappa\lambda} : \hat{e}^{uv}_{cd}|F\rangle = \Delta^{uv}_{rs} \langle F|\hat{b}^+_a \hat{b}^+_b : \hat{b}^+_\mu \hat{b}^+_\nu \hat{b}_\lambda \hat{b}_\kappa : \hat{b}_d \hat{b}_c|F\rangle +$$
$$+ \Delta^{ab}_{cd} \langle F|\hat{b}_s \hat{b}_r : \hat{b}^+_\mu \hat{b}^+_\nu \hat{b}_\lambda \hat{b}_\kappa : \hat{b}^+_u \hat{b}^+_v|F\rangle +$$
$$+ \mathcal{A}^{\mu\nu}_{\kappa\lambda} \mathcal{A}^{ab}_{cd} \delta^\nu_d \delta^b_\kappa \langle F|\hat{b}^+_a \hat{b}_s \hat{b}_r : \hat{b}^+_\mu \hat{b}_\lambda : \hat{b}^+_u \hat{b}^+_v \hat{b}_c|F\rangle.$$

Clearly, in the first term we contracted four particle operators, in the second term we contracted four hole operators, and in the last term we contracted two particle

3.4 Coupled Cluster Equations in Spin-orbital Form

and two hole operators. The remaining vacuum averages are

$$\langle F|\hat{b}_a^+\hat{b}_b^+ : \hat{b}_\mu^+\hat{b}_\nu^+\hat{b}_\lambda\hat{b}_\kappa : \hat{b}_d\hat{b}_c|F\rangle = \Delta_{\kappa\lambda}^{ab}\Delta_{cd}^{\mu\nu},$$

$$\langle F|\hat{b}_s\hat{b}_r : \hat{b}_\mu^+\hat{b}_\nu^+\hat{b}_\lambda\hat{b}_\kappa : \hat{b}_u^+\hat{b}_v^+|F\rangle = \Delta_{rs}^{\mu\nu}\Delta_{\kappa\lambda}^{uv}$$

and

$$\langle F|\hat{b}_a^+\hat{b}_s\hat{b}_r : \hat{b}_\mu^+\hat{b}_\lambda : \hat{b}_u^+\hat{b}_v^+\hat{b}_c|F\rangle = \delta_c^a \mathcal{A}_{rs}^{uv}\delta_r^\mu\delta_\lambda^u\delta_s^v.$$

Substituting the last four equations into the second term on the rhs of Eq. (3.46), we obtain

$$\frac{1}{4}v_{\mu\nu}^{\kappa\lambda}\langle F|\hat{e}_{rs}^{ab} : \hat{e}_{cd}^{\mu\nu} : \hat{e}_{\kappa\lambda}^{uv}|F\rangle = v_{cd}^{ab}\Delta_{rs}^{uv} + v_{rs}^{uv}\Delta_{cd}^{ab} + \mathcal{A}_{rs}^{ab}\mathcal{A}_{cd}^{uv}v_{rc}^{au}\delta_s^v\delta_d^b, \quad (3.48)$$

where we used Eq. (3.45). By substituting Eqs. (3.47) and (3.48) into Eq. (3.46) we obtain the rhs of Eq. (3.41).

To derive Eq. (3.42) we again substitute Eq. (3.38) for the normally ordered Hamiltonian into its lhs,

$$\langle F|\hat{e}_{rs}^{ab}\left[-2\hat{e}_{cd}^{uv} : \hat{H} : + : \hat{H} : \hat{e}_{cd}^{uv}\right]\hat{e}_{ef}^{xy}|F\rangle =$$
$$= \frac{1}{4}v_{\mu\nu}^{\kappa\lambda}\langle F|\hat{e}_{rs}^{ab}\left[-2\hat{e}_{cd}^{uv} : \hat{e}_{\kappa\lambda}^{\mu\nu} : + : \hat{e}_{\kappa\lambda}^{\mu\nu} : \hat{e}_{cd}^{uv}\right]\hat{e}_{ef}^{xy}|F\rangle. \quad (3.49)$$

The vacuum averages to be evaluated are

$$\langle F|\hat{e}_{rs}^{ab} : \hat{e}_{\kappa\lambda}^{\mu\nu} : \hat{e}_{cd}^{uv}\hat{e}_{ef}^{xy}|F\rangle =$$
$$= \langle F|\hat{b}_a^+\hat{b}_b^+\hat{b}_\mu^+\hat{b}_\nu^+\hat{b}_d\hat{b}_c\hat{b}_f\hat{b}_e|F\rangle\langle F|\hat{b}_s\hat{b}_r\hat{b}_\lambda\hat{b}_\kappa\hat{b}_u^+\hat{b}_v^+\hat{b}_x^+\hat{b}_y^+|F\rangle = \Delta_{efcd}^{ab\mu\nu}\Delta_{\kappa\lambda rs}^{uvxy}$$

and

$$\langle F|\hat{e}_{rs}^{ab}\hat{e}_{cd}^{uv} : \hat{e}_{\kappa\lambda}^{\mu\nu} : \hat{e}_{ef}^{xy}|F\rangle = \Delta_{cd}^{ab}\Delta_{rs}^{uv}\Delta_{\kappa\lambda}^{xy}\Delta_{ef}^{\mu\nu}.$$

Here, the symbol $\Delta_{efcd}^{ab\mu\nu}$ is completely antisymmetric in the upper and lower indices, a, b, μ, ν and e, f, c, d, respectively. It comprises 4!=24 terms; they can be most suitably expressed as

$$\Delta_{efcd}^{ab\mu\nu} = \Delta_{ef}^{ab}\Delta_{cd}^{\mu\nu} + \Delta_{cd}^{ab}\Delta_{ef}^{\mu\nu} - \mathcal{A}_{cd}\mathcal{A}_{ef}\Delta_{ec}^{ab}\Delta_{fd}^{\mu\nu}.$$

Let us check: the rhs comprises $2 \times 2 + 2 \times 2 + 2 \times 2 \times 2 \times 2 = 4 + 4 + 16 = 24$ terms. Similarly, the symbol $\Delta_{\kappa\lambda rs}^{uvxy}$ can be most suitably expressed as

$$\Delta_{\kappa\lambda rs}^{uvxy} = \Delta_{\kappa\lambda}^{uv}\Delta_{rs}^{xy} + \Delta_{rs}^{uv}\Delta_{\kappa\lambda}^{xy} - \mathcal{A}^{uv}\mathcal{A}^{xy}\Delta_{\kappa\lambda}^{ux}\Delta_{rs}^{vy}. \tag{3.50}$$

Substituting now the last four equations into Eq. (3.49) and using the fact that expression (3.49) is multiplied by the amplitudes $t_{cd}^{uv}t_{ef}^{xy}$, see Eq. (3.40), so that the terms that differ by simultaneous interchange $(uv) \leftrightarrow (xy)$ and $(cd) \leftrightarrow (ef)$ contribute equally, we obtain Eq. (3.42). Note that the expression for $\Delta_{efcd}^{ab\mu\nu}$ and $\Delta_{\kappa\lambda rs}^{uvxy}$ are manifestly symmetric with respect to the interchange $(cd) \leftrightarrow (ef)$ and $(uv) \leftrightarrow (xy)$, respectively.

Finally, it is worth noting that the Eq. (3.40) are manifestly antisymmetric in the indices a and b and in the indices r and s, cf. Eqs. (3.45), (3.41) and (3.42). This is not surprising as the excitation operator $\hat{e}_{ab}^{rs}$ is antisymmetric in these indices, cf. Eq. (1.38). Likewise, the expressions (3.41) and (3.42) are antisymmetric in the indices u and v, the indices c and d, the indices x and y and the indices e and f. These reflect the antisymmetry of the pertinent excitations operators.

Chapter 4
Further Developments

If the Hamiltonian is non-relativistic, see, e.g., Eqs. (1.1)–(1.3), then it is spin-independent. Consequently, the total spin of the electrons is preserved and one can search for common eigenvectors of the operators $\hat{H}, \hat{S}^2, \hat{S}_z$, where $\hat{S}^2$ and $\hat{S}_z$ designate the square and the projection on the z-axis of the total spin operator. Thus, in addition to the permutational symmetry, we include the spin symmetry as well. This naturally begs the question how to use these two symmetries to simplify the approximate calculation.[1] Further, we show how the system of quadratic equations for the cluster amplitudes can be solved by the iterative Newton-Raphson method. We illustrate all these points by applying the cc method to the Hubbard model of benzene. Further, we show how to include monexcitations exactly and triexcitations perturbatively in a cc calculation. We conclude this chapter by showing how one can apply the cc method to the one-electron open-shell systems.

4.1 Adaptation to Permutational Symmetry

There is an additional simplification that can be traced back to the Pauli exclusion principle, that is, to the antisymmetry of the matrix elements (3.45) and cluster amplitudes, cf. Eq. (3.29),

$$t_{ab}^{rs} = -t_{ab}^{sr} = -t_{ba}^{rs} = t_{ba}^{sr}. \qquad (4.1)$$

---

[1] We proceed here in a very direct manner. Over the years, physicists developed a number of strategies to combine the permutational and spin symmetries. See for instance [19] and [7] for a survey and an exposition of the existing approaches in molecular and atomic cases, respectively.

Equations (3.40) hold for unordered pairs of upper and lower indices, r, s and a, b, respectively. For given pairs of upper and lower indices, one can form four orthogonal linear combinations of unordered amplitudes that differ by the (anti)symmetry with respect to the exchange of upper and lower indices

$$\begin{pmatrix} \left(t_{\overline{ab}}^{\overline{rs}}\right)^{++} \\ \left(t_{\overline{ab}}^{\overline{rs}}\right)^{+-} \\ \left(t_{\overline{ab}}^{\overline{rs}}\right)^{-+} \\ \left(t_{\overline{ab}}^{\overline{rs}}\right)^{--} \end{pmatrix} = \frac{1}{2} \begin{pmatrix} 1 & 1 & 1 & 1 \\ 1 & 1 & -1 & -1 \\ 1 & -1 & 1 & -1 \\ 1 & -1 & -1 & 1 \end{pmatrix} \begin{pmatrix} t_{ab}^{rs} \\ t_{ab}^{sr} \\ t_{ba}^{rs} \\ t_{ba}^{sr} \end{pmatrix}. \tag{4.2}$$

This transformation is orthogonal, so the inverse transformation reads

$$\begin{pmatrix} t_{ab}^{rs} \\ t_{ab}^{sr} \\ t_{ba}^{rs} \\ t_{ba}^{sr} \end{pmatrix} = \frac{1}{2} \begin{pmatrix} 1 & 1 & 1 & 1 \\ 1 & 1 & -1 & -1 \\ 1 & -1 & 1 & -1 \\ 1 & -1 & -1 & 1 \end{pmatrix} \begin{pmatrix} \left(t_{\overline{ab}}^{\overline{rs}}\right)^{++} \\ \left(t_{\overline{ab}}^{\overline{rs}}\right)^{+-} \\ \left(t_{\overline{ab}}^{\overline{rs}}\right)^{-+} \\ \left(t_{\overline{ab}}^{\overline{rs}}\right)^{--} \end{pmatrix}. \tag{4.3}$$

Now, if we substitute the last equation into Eq. (3.43) and Eqs. (3.40)–(3.42), we find that, due to the antisymmetry of the potential matrix elements (3.45), only the combination $\left(t_{\overline{ab}}^{\overline{rs}}\right)^{--}$ contributes to the correlation energy and this combination decouples from the other three combinations. Thus, the last equation can be simplified to

$$t_{ab}^{rs} = \frac{1}{2} \mathcal{A}_{ab}^{rs} T_{\overline{ab}}^{\overline{rs}}, \quad T_{\overline{ab}}^{\overline{rs}} = \left(t_{\overline{ab}}^{\overline{rs}}\right)^{--}. \tag{4.4}$$

Further, we order not only the amplitudes, but Eq. (3.40) themselves,

$$E_{\overline{rs}}^{\overline{ab}} = \frac{1}{2} \mathcal{A}_{rs}^{ab} e_{rs}^{ab} = 2v_{rs}^{ab} + T_{\overline{cd}}^{\overline{uv}} \mathcal{L}_{rs,cd}^{ab,uv} + \frac{1}{2^2} T_{\overline{cd}}^{\overline{uv}} T_{\overline{ef}}^{\overline{xy}} \mathcal{Q}_{rs,cd,ef}^{ab,uv,xy} = 0, \tag{4.5}$$

where we substituted Eq. (4.4) and used the fact that the expressions (3.40), (3.41) and (3.42) are automatically antisymmetric with respect to the interchange of the pertinent indices and each additional antisymmetrization in each pair of indices produces multiplication of the expression by 2. Finally, substituting Eq. (4.4) into Eq. (3.43), we obtain the correlation energy in terms of the ordered amplitudes

$$E_{\text{corr}} = \frac{1}{2} v_{rs}^{ab} T_{\overline{ab}}^{\overline{rs}}. \tag{4.6}$$

4.2 Adaptation to Spin Symmetry

Molecules in general do not possess any spatial symmetry. In most of the cases of interest, it suffices to consider a non-relativistic approximation, see, e.g., Eqs. (1.1), (1.2) and (1.3). The Hamiltonian is spin-independent and commutes with the components of the spin of the electron field, Eq. (2.41). Atoms possess spherical symmetry, the Hamiltonian, whether relativistic or non-relativistic, commutes with total angular momentum operator of the electron field

$$\hat{\mathbf{J}} = J_\mu^\nu \hat{\mathbf{e}}_\nu^\mu ,$$

where J_μ^ν are the matrix elements of one-electron total angular momentum operator; explicitly

$$J_\mu^\nu = \langle \mu | \hat{\mathbf{L}} + \frac{\boldsymbol{\Sigma}}{2} | \nu \rangle .$$

Here, the first and second operators inside the inner product apparently stand for the one-electron orbital angular momentum and spin operators, respectively. Also the Hamiltonian, whether relativistic or non-relativistic, is invariant with respect to inversion of coordinate axes. This leads to parity conservation. Thus, we shall be more general than in Eq. (2.43) and assume the restricted Hartree-Fock method yields the spin-orbitals in the form

$$|a\rangle = |A\rangle |j_a, m_a, \kappa_a\rangle ,$$

where $j_a(j_a + 1)$ and $m_a = -j_a, -j_a + 1, \ldots, j_a - 1, j_a$ designate the spin-orbital's square of the total angular momentum, and its projection to one of the coordinate axes, respectively and $(-1)^{j_a - \kappa_a/2}$ is the parity of the state, $\kappa_a = \pm 1$. In the case of molecules $|A\rangle$ designates a molecular orbital and $j_a = \frac{1}{2}$. If the nuclear configuration is not invariant to inversion of coordinate axes, then κ is ignored. In the case of atoms $|A\rangle$ designates a radial part of an atomic orbital and j_a takes any half-integer value. In the following we shall refer to $|A\rangle$'s as orbitals.

Considering then the following linear combination of the matrix elements (1.39)

$$\sum_{m_a=-j_a}^{j_a} \sum_{m_r=-j_r}^{j_r} (j_a, m_a, j_b, m - m_a | j)(j_r, m_r, j_s, m' - m_r | j') v_{rs}^{ab} = \quad (4.7)$$

$$= v_{AB}^{RS}(j) \delta_{m,m'} \delta_{j',j}$$

we find the result (rhs of the last equation) is non-zero only if the resulting angular momenta j, j', and their projections m, m', of holes and particles are the same. If this is the case, the result is independent of the total magnetic quantum number m.

Owing to the reality and orthonormality of Clebsch-Gordan (CG) coefficients, the potential matrix elements (1.39) can be expressed in terms of symmetry adapted elements as

$$v_{rs}^{ab} = \sum_{j=\max(|j_a-j_b|,|j_r-j_s|)}^{\min(j_a+j_b,j_r+j_s)} (j_a, m_a, j_b, m - m_a|j)(j_r, m_r, j_s, m - m_r|j) v_{AB}^{RS}(j). \tag{4.8}$$

Finally, substituting Eq. (4.8) into Eq. (3.45) we obtain by virtue of the symmetry of the CG coefficients

$$v_{AB}^{RS}(j) = (-1)^{j-j_a-j_b-1} v_{BA}^{RS}(j) = (-1)^{j-j_r-j_s-1} v_{AB}^{SR}(j). \tag{4.9}$$

This in particular yields

$$v_{AA}^{RS}(j) = (-1)^{j-2j_a-1} v_{AA}^{RS}(j) = (-1)^j v_{AA}^{RS}(j), \tag{4.10}$$

where in the last equation we took the advantage of the fact that since j_a is a half-integer, $2j_a + 1$ is even. Whence the matrix elements $v_{AA}^{RS}(j)$ vanish for odd j. Further, for one-electron states the reflection of coordinate axes around origin produces the factor $(-1)^{j-\kappa/2}$. Whence the matrix elements v_{rs}^{ab} and $v_{AB}^{RS}(j)$ vanish unless the combined parities of holes and particles are the same

$$(-1)^{j_a+j_b-(\kappa_a+\kappa_b)/2} = (-1)^{j_r+j_s-(\kappa_r+\kappa_s)/2}. \tag{4.11}$$

We can make an orthogonal transformation between the *unordered* amplitudes t_{ab}^{rs} and amplitudes $t_{AB}^{RS}(j_+, j_-, J)$ describing the biexcited states of definite total angular momenta

$$t_{ab}^{rs} = \sum_{j_+,j_-,J} (j_a, m_a, j_b, m_b|j_-)(j_r, m_r, j_s, m_s|j_+) \times$$

$$\times (j_+, m_r + m_s, j_-, -m_a - m_b|J)(-1)^{j_- -m_a -m_b} t_{AB}^{RS}(j_+, j_-, J).$$

Clearly, we first add the spins of particles to get the total particle spin j_+ and the spins of holes to get the total hole spin j_- and then combine particle and hole spins to get the total spin J. This is usually referred to as pp-hh coupling scheme. When adding the spins of particles and holes, the magnetic quantum numbers of holes have to be taken with opposite sign and the pertinent CG coefficient has to be multiplied by the appropriate phase factor, see discussion in Sect. 2.5.

4.2 Adaptation to Spin Symmetry

Now, due to spin symmetry of the Hamiltonian, the amplitudes $t_{AB}^{RS}(j_+, j_-, J)$ with different J decouple from each other. If we are interested in the ground state energy, as is usually the case, it suffices to keep only the term with $J = 0$ in the expansion on the rhs in the last equation, namely,

$$t_{ab}^{rs} = \sum_{j=\max(|j_a-j_b|,|j_r-j_s|)}^{\min(j_a+j_b, j_r+j_s)} (j_a, m_a, j_b, m_b|j)(j_r, m_r, j_s, m_s|j)\delta_{m_a+m_b}^{m_r+m_s} t_{AB}^{RS}(j), \quad (4.12)$$

where

$$t_{AB}^{RS}(j) = \frac{t_{AB}^{RS}(j, j, 0)}{\sqrt{2j+1}}.$$

When considering the above transformation between *ordered* amplitudes we have to distinguish if the orbitals are the same or not, so Eq. (4.12) is modified as follows:

$$T_{\overline{ab}}^{\overline{rs}} = \sum_{j=\max(|j_a-j_b|,|j_r-j_s|)}^{\min(j_a+j_b, j_r+j_s)} D_{j_a,m_a,j_b,m_b,\delta_{AB}}^{j_r,m_r,j_s,m_s,\delta_{RS}}(j) T_{\overline{AB}}^{\overline{RS}}(j), \quad (4.13)$$

where

$$D_{j_a,m_a,j_b,m_b,\delta_{AB}}^{j_r,m_r,j_s,m_s,\delta_{RS}}(j) = V_{j_a,j_b,\delta_{AB}}^{j_r,j_s,\delta_{RS}}(j)(j_a, m_a, j_b, m_b|j)(j_r, m_r, j_s, m_s|j)\delta_{m_a+m_b}^{m_r+m_s} \times \quad (4.14)$$

$$\times \left[\delta_{AB}(1-(-1)^{j-j_a-j_b})\Theta(m_b - m_a) + 1 - \delta_{AB}\right] \times$$

$$\times \left[\delta_{RS}(1-(-1)^{j-j_r-j_s})\Theta(m_s - m_r) + 1 - \delta_{RS}\right],$$

where $\Theta(x)$ is Heaviside function, $\Theta(x) = 1$ for $x > 0$ and zero otherwise. Further, $V_{j_a,j_b,\delta_{AB}}^{j_r,j_s,\delta_{RS}}(j)$ is determined from the normalization condition

$$\sum_{m_a,m_b,m_r,m_s} \left[D_{j_a,m_a,j_b,m_b,\delta_{AB}}^{j_r,m_r,j_s,m_s,\delta_{RS}}(j)\right]^2 = 1. \quad (4.15)$$

If the orbitals A and B, and R and S are different, then the coefficients D are the same as those in Eq. (4.12). But if, say the orbitals A and B, are the same, then, firstly, the requirement of the ordering enforces the restriction $m_a < m_b$. Secondly, we sum over the permutation of the spin-orbitals a and b to exclude the values of j that are symmetric with respect of the interchange of the spin-orbitals a and b, since then the Pauli exclusion principle is not satisfied. Note that these two changes do not spoil the orthogonality of the transformation (4.12). For instance, if the orbitals

A and B are the same, then $j_a = j_b = j$. The orthogonality relations for CG coefficients then read

$$\sum_{m=-j}^{j} (j, -m, j, m|J)(j, -m, j, m|J') = \delta_{J,J'}.$$

Now, for half-integral j this can be rewritten as

$$\sum_{m=-j}^{-1/2} (j, -m, j, m|J)(j, -m, j, m|J') \left[1 + (-1)^{J+J'-4j}\right] = \delta_{J,J'},$$

where the term in the square bracket is apparently independent of m. Since j is half-integral, the term in the square-bracket vanishes if the sum $J + J'$ is odd. If the sum $J + J'$ is even, then apparently

$$\sum_{m=-j}^{-1/2} (j, -m, j, m|J)(j, -m, j, m|J') = \frac{\delta_{J,J'}}{2}.$$

Thus, the orthogonality is preserved and only the normalization has to be taken care of. Finally, due to parity conservation, the cluster amplitudes $T_{\overline{AB}}^{\overline{RS}}(j)$ vanish unless condition (4.11) holds.

Substituting now the transformation (4.13) into Eq. (4.5) we obtain coupled-cluster equations for the ordered symmetry adapted amplitudes $T_{\overline{AB}}^{\overline{RS}}(j)$

$$E_{\overline{RS}}^{\overline{AB}}(J) = \sum_{m_a,m_b,m_r,m_s} D_{j_r,m_r,j_s,m_s,\delta_{RS}}^{j_a,m_a,j_b,m_b,\delta_{AB}}(J) E_{\overline{rs}}^{\overline{ab}}. \qquad (4.16)$$

The explicit form of these equations is rather lengthy. It is not necessary to give it here, as the construction of these equations is very simple. Let us illustrate it on the symmetry adaptation of the last term in Eq. (3.41),

$$\mathcal{A}_{rs}^{ab} \mathcal{A}_{cd}^{uv} v_{rc}^{au} \delta_v^s \delta_b^d \rightarrow \qquad (4.17)$$

$$\rightarrow \sum_{J''} \sum_{p_1,p_2,q_1,q_2} \prod_{i=1}^{2} \text{sgn}(\mathcal{P}_{p_i})\text{sgn}(\mathcal{P}_{q_i}) [S]_{j_r,j_s,\delta_{RS},p_1,j_c,j_d,\delta_{CD},p_2}^{j_a,j_b,\delta_{AB},q_1,j_u,j_v,\delta_{UV},q_2}(J,J',J'') \times$$

$$\times v_{\mathcal{P}_{p_1}(R),\mathcal{P}_{p_2}(C)}^{\mathcal{P}_{q_1}(A),\mathcal{P}_{q_2}(U)}(J'') \delta_{\mathcal{P}_{q_2}(V)}^{\mathcal{P}_{p_1}(S)} \delta_{\mathcal{P}_{q_1}(B)}^{\mathcal{P}_{p_2}(D)},$$

where the spinor factor in this case reads

$$[S]^{j_a,j_b,\delta_{AB},q_1,j_u,j_v,\delta_{UV},q_2}_{j_r,j_s,\delta_{RS},p_1,j_c,j_d,\delta_{CD},p_2}(J,J',J'') = \sum_{\text{all } m's} D^{j_a,m_a,j_b,m_b,\delta_{AB}}_{j_r,m_r,j_s,m_s,\delta_{RS}}(J) \times \quad (4.18)$$

$$\times D^{j_u,m_u,j_v,m_v,\delta_{UV}}_{j_c,m_c,j_d,m_d,\delta_{CD}}(J')(j_{\mathcal{P}_{p_1}(r)}, m_{\mathcal{P}_{p_1}(r)}, j_{\mathcal{P}_{p_2}(c)}, m_{\mathcal{P}_{p_2}(c)}|J'') \times$$

$$\times (j_{\mathcal{P}_{q_1}(a)}, m_{\mathcal{P}_{q_1}(a)}, j_{\mathcal{P}_{q_2}(u)}, m_{\mathcal{P}_{q_2}(u)}|J'')\delta^{m_{\mathcal{P}_{q_1}(a)}+m_{\mathcal{P}_{q_2}(u)}}_{m_{\mathcal{P}_{p_1}(r)}+m_{\mathcal{P}_{p_2}(c)}} \delta^{m_{\mathcal{P}_{p_1}(s)}}_{m_{\mathcal{P}_{q_2}(v)}} \delta^{m_{\mathcal{P}_{p_2}(d)}}_{m_{\mathcal{P}_{q_1}(b)}}.$$

The permutations are defined as, for instance,

$$\mathcal{P}_1(a) = a, \quad \mathcal{P}_1(b) = b, \quad \text{sgn}(\mathcal{P}_1) = 1,$$

$$\mathcal{P}_2(a) = b, \quad \mathcal{P}_2(b) = a, \quad \text{sgn}(\mathcal{P}_2) = -1$$

and so on. It is seen from Eq. (4.17) that symmetry adaptation is achieved by replacing the pertinent spin-orbitals with orbitals. The expression has to be, for given permutation, multiplied by appropriate spinor factor. This spinor factor is obtained, for given permutation, by replacing the matrix elements of the Coulomb interaction and cluster amplitudes by the appropriate CG coefficients and their generalization, cf. Eqs. (4.8) and (4.13), respectively. In more detail, the first term in the spinor factor (4.18), $D^{j_a,m_a,j_b,m_b,\delta_{AB}}_{j_r,m_r,j_s,m_s,\delta_{RS}}(J)$, comes from Eq. (4.16). The term on lhs of Eq. (4.17) is multiplied by the ordered amplitude $T^{\overline{uv}}_{cd}$, see Eq. (4.5), so the second term in the spinor factor (4.18), $D^{j_u,m_u,j_v,m_v,\delta_{UV}}_{j_c,m_c,j_d,m_d,\delta_{CD}}(J')$, comes from substitution (4.13). The third, fourth and fifth terms in the spinor factor (4.18), the two CG coefficients and Kronecker delta $\delta^{m_{\mathcal{P}_{q_1}(a)}+m_{\mathcal{P}_{q_2}(u)}}_{m_{\mathcal{P}_{p_1}(r)}+m_{\mathcal{P}_{p_2}(c)}}$ come from substitution (4.8) for potential matrix elements. The last two Kronecker delta's in the spinor factor (4.18) come from substitution $\delta^s_v \to \delta^{\mathcal{P}_{p_1}(S)}_{\mathcal{P}_{q_2}(V)}\delta^{m_{\mathcal{P}_{p_1}(s)}}_{m_{\mathcal{P}_{q_2}(v)}}$ and analogous substitution for δ^d_b.

4.3 Perturbative Solution of Coupled Cluster Equations

Coupled cluster equations (3.40) can be solved in the perturbative manner, meaning we expand the cluster amplitudes according to powers of the interelectronic potential v,

$$t^{rs}_{ab} = (t^{rs}_{ab})^{(1)} + (t^{rs}_{ab})^{(2)} + \ldots,$$

where we have from Eqs. (3.40) and (3.41)

$$0 = v_{rs}^{ab} + (\epsilon_r + \epsilon_s - \epsilon_a - \epsilon_b)(t_{ab}^{rs})^{(1)},$$

$$0 = (\epsilon_r + \epsilon_s - \epsilon_a - \epsilon_b)(t_{ab}^{rs})^{(2)} + \frac{1}{2^2}(t_{cd}^{uv})^{(1)} \left[v_{cd}^{ab}\Delta_{rs}^{uv} + v_{rs}^{uv}\Delta_{cd}^{ab} + \mathcal{A}_{rs}^{ab}\mathcal{A}_{cd}^{uv} v_{rc}^{au} \delta_s^v \delta_d^b \right],$$

and so on. By inserting this into Eq. (3.43) we get perturbative expansion of correlation energy[2]

$$E_{\text{MP2}} = \frac{1}{4} v_{rs}^{ab}(t_{ab}^{rs})^{(1)} = -\frac{1}{4}\frac{v_{rs}^{ab} v_{ab}^{rs}}{\epsilon_r + \epsilon_s - \epsilon_a - \epsilon_b} \quad (4.19)$$

$$E_{\text{MP3}} = \frac{1}{4} v_{rs}^{ab}(t_{ab}^{rs})^{(2)}$$

and so on. Actually, the fourth order perturbative energy obtained from Eqs. (3.40) and (3.43) will differ from the correct fourth order perturbative energy by missing contribution from the cluster operators $\hat{T}_1$, $\hat{T}_3$ and $\hat{T}_4$. These considerations are of more theoretical than practical significance. The perturbation method is rarely used beyond the second order, Eq. (4.19), as the convergence of the perturbation expansion is known to be slow.

4.4 Iterative Solution of Coupled Cluster Equations

Adaptation to permutational and spin symmetries described above significantly reduces the number of biexcitations. After this is done we have to make unique assignment between natural numbers and ordered spin adapted biexcitations

$$n \leftrightarrow \left(\overline{\frac{RS}{AB}}\right)(J).$$

When this is done, Eq. (4.16) present a system of quadratic equations

$$A_i + L_{i,j} t_j + Q_{i,j,k} t_j t_k = 0, \quad (4.20)$$

---

[2] The subscripts MP stands for Christian Moller (1904–1980) and Milton S. Plesset (1908–1991), who were first to consider the perturbative solution of Eq. (3.9).

where all indices i, j, k run over all possible ordered spin adapted biexcitations. We search for solution of this system by Newton-Raphson iterative method. At the zeroth order approximation we neglect quadratic terms in Eq. (4.20)

$$L_{i,j} t_j^{(0)} = -A_i .\tag{4.21}$$

Further, we write the $(l + 1)$-th iteration as

$$t_j^{(l+1)} = t_j^{(l)} + \delta_j^{(l)} ,\tag{4.22}$$

plug it into Eq. (4.20) and neglect quadratic terms in δ's:

$$[L_{i,j} + (Q_{i,j,k} + Q_{i,k,j}) t_k^{(l)}] \delta_j^{(l)} = - \left(A_i + L_{i,j} t_j^{(l)} + Q_{i,j,k} t_j^{(l)} t_k^{(l)} \right) .\tag{4.23}$$

We continue this process until we reach self-consistency, that is until the corresponding correlation energies do not significantly differ. Equations (4.21) and (4.23) present systems of linear equations that can be solved using standard linear algebra packages.

4.5 Hubbard Model of Benzene

Hubbard model of benzene is the simplest model of electron correlation that bears some resemblance to the reality. It is well-known, see e.g. [20, 21], in a good approximation the π-electrons move independently on the σ-electrons. Restricting ourselves to the π-electrons we assume the one- and two-particle matrix elements, Eqs. (1.31) and (1.39), read

$$z_\nu^\mu = -t \delta_{N \pm 1}^M \delta_n^m ,\tag{4.24}$$

and

$$v_{\mu\nu}^{\kappa\lambda} = v \delta_K^M \delta_L^N \delta_L^M \Delta_{kl}^{mn}\tag{4.25}$$

where, cf. Eq. (2.43),

$$|\mu\rangle = |M\rangle |m\rangle , |\nu\rangle = |N\rangle |n\rangle , \langle\kappa| = \langle K| \langle k| , \langle\lambda| = \langle L| \langle l| .$$

If we ignore the two-particle term, the one-particle term describes what is known as Hückel model, see e.g. [21]. The π-electrons move independently on each other. They can jump from each of the six carbon atoms to two neighbouring carbon atoms. The Hubbard interaction term, Eq. (4.25), describes interaction of the π-electrons sitting on the same carbon atom, all other interactions are neglected. The realistic values of the parameters are $t = 2.5\,\text{eV}$ and $v = 5\,\text{eV}$ [22].

Searching for the solution of the one-particle Schrödinger equation

$$\hat{z}|\psi_k\rangle = \epsilon_k |\psi_k\rangle \tag{4.26}$$

in the form of the linear combination of the atomic orbitals

$$|\psi_k\rangle = \sum_{A=1}^{6} c_A^k |A\rangle, \tag{4.27}$$

where $|A\rangle$ describes the atomic π-orbital localized on the A-th carbon atom, and multiplying Eq. (4.26) successively by vectors $\langle B|$, $B = 1, \ldots, 6$ we obtain the matrix eigenvalue problem

$$z_B^A c_A^k = \langle B|A\rangle \epsilon_k c_A^k \Rightarrow -t\delta_{B\pm 1}^A c_A^k = \epsilon_k c_B^k, \tag{4.28}$$

where we used Eq. (4.24) and assumed the orthonormality of the atomic orbitals, $\langle B|A\rangle = \delta_B^A$. Searching for the solution of Eq. (4.28) in the form $c_A^k = q^A$, subject to the boundary condition $c_7^k = c_1^k$ (carbons form closed ring) and normalization condition $\langle \psi | \psi \rangle = 1$, we obtain

$$\epsilon_k = -2t\cos(\pi k/3), \quad c_A^k = \frac{1}{\sqrt{6}} \exp\left\{\frac{i\pi k A}{3}\right\}, \quad k = 0, \ldots, 5. \tag{4.29}$$

The values of these energies are explicitly displayed in the Table 4.1. It is seen that the occupied orbitals belong to $k = 0, 1, 5$ and virtual orbitals belong to $k = 2, 3, 4$.

As can be easily verified the Hartree-Fock orbitals are identical to these Hückel orbitals. The matrix elements of the Fock operator are given by, cf. Eq. (2.29),

$$\langle \psi_\mu | \hat{f} | \psi_\nu \rangle = \langle \psi_\mu | \hat{z} | \psi_\nu \rangle + \sum_{b=0,1,5} \langle \psi_\mu |_1 \langle \psi_b |_2 \hat{r}_{12}^{-1} \left(2|\psi_\nu\rangle_1 |\psi_b\rangle_2 - |\psi_b\rangle_1 |\psi_\nu\rangle_2\right).$$

Table 4.1 One-particle Hartree-Fock energies for Hubbard model of benzene

k	ϵ_k/t
0	-2
1	-1
2	1
3	2
4	1
5	-1

4.5 Hubbard Model of Benzene

Substituting Eqs. (4.27) and (4.29) into the last equation one obtains

$$\langle\psi_\mu|\hat{f}|\psi_\nu\rangle = \frac{1}{6}\sum_{P,Q}\exp\left\{-i\frac{\pi}{3}(\mu P - \nu Q)\right\}\langle P|\hat{z}|Q\rangle+$$

$$+\frac{1}{6^2}\sum_{b,P,Q,R,S}\exp\left\{-i\frac{\pi}{3}[\mu P - \nu Q + b(R-S)]\right\}\times$$

$$\times\langle P|_1\langle R|_2\hat{r}_{12}^{-1}(2|Q\rangle_1|S\rangle_2 - |S\rangle_1|Q\rangle_2) =$$

$$= \frac{1}{6}\sum_{P=1}^{6}\exp\left\{-i\frac{\pi}{3}(\mu-\nu)P\right\}\left[-2t\cos\left(\frac{\pi}{3}\nu\right)+\frac{v}{2}\right] = \delta_\mu^\nu\left[-2t\cos\left(\frac{\pi}{3}\nu\right)+\frac{v}{2}\right],$$

where in the second equality we substituted from Eqs. (4.24) and (4.25). The Fock matrix is in the Hückel basis diagonal. The Hartree-Fock one-particle energies are shifted from Hückel energies by inessential constant $v/2$.

The two-particle matrix elements in the Hartree-Fock-Hückel basis then reads

$$v_{jn}^{kl} = v\Delta_{m_j,m_n}^{m_k,m_l}\frac{1}{6^2}\sum_{A=1}^{6}\exp\left\{\frac{i\pi A(k+l-j-n)}{3}\right\}$$

$$= v\Delta_{m_j,m_n}^{m_k,m_l}\frac{\delta_{j+n+6q}^{k+l}}{6}, \quad q = 0, \pm 1, \pm 2, \ldots, \quad (4.30)$$

where we substituted for the spin-orbitals, cf. Eq. (2.43),

$$|k\rangle = |\psi_k\rangle|m_k\rangle, |l\rangle = |\psi_l\rangle|m_l\rangle, \langle j| = \langle\psi_j|\langle m_j|, \langle n| = \langle\psi_n|\langle m_n|,$$

where the orbitals $|\psi_k\rangle$ are given by Eqs. (4.27) and (4.29) and where we used the two-particle matrix elements in the basis of the atomic orbitals, Eq. (4.25).

Usefulness of adaptation of coupled cluster equations to permutational and spin symmetries described above can now be readily illustrated. Consider for example biexcitation from orbitals $k = 1, 5$ to orbitals $2, 4$. When considering coupled cluster equations in spin-orbital form, Eq. (3.40), we have to consider all possible 2^2 spin configurations: we consider either all upper or all lower signs of spins

$$\begin{pmatrix}(2,\pm) & (4,\pm)\\(1,\pm) & (5,\pm)\end{pmatrix}$$

and

$$\begin{pmatrix}(2,\pm) & (4,\mp)\\(1,\pm) & (5,\mp)\end{pmatrix}.$$

The spin configurations where the sum of the spins in the virtual spin-orbitals differs from the sum of the spins in the occupied spin-orbitals are excluded because of conservation of total $\hat{S}_z$. Further, for each spin configuration we have to consider all possible orderings, e.g.

$$\begin{pmatrix} (2, \pm) \ (4, \pm) \\ (1, \pm) \ (5, \pm) \end{pmatrix},$$

$$\begin{pmatrix} (4, \pm) \ (2, \pm) \\ (1, \pm) \ (5, \pm) \end{pmatrix},$$

$$\begin{pmatrix} (2, \pm) \ (4, \pm) \\ (5, \pm) \ (1, \pm) \end{pmatrix}$$

and

$$\begin{pmatrix} (4, \pm) \ (2, \pm) \\ (5, \pm) \ (1, \pm) \end{pmatrix}.$$

With adaptation to permutational and spin symmetries instead of these 2^4 configurations we have to consider just 2 configurations

$$\begin{pmatrix} 4 & 2 \\ 5 & 1 \end{pmatrix} (0)$$

and

$$\begin{pmatrix} 4 & 2 \\ 5 & 1 \end{pmatrix} (1).$$

In the Table 4.2 the convergence of Newton-Raphson method for solution of coupled cluster equations is illustrated.

Comparing this result to the exact solution of Hubbard model $E_{\text{corr}}/t \simeq -0.40944$ [22] we see the coupled cluster method works fairly well. Note that in this case the second-order perturbation method, Eq. (4.19), yields $E_{\text{MP2}}/t = -0.402777\ldots$.

Table 4.2 Iterations of coupled cluster equations for Hubbard model of benzene

Iteration	E_{corr}/t
0	−0.422199121
1	−0.408969747
2	−0.408955910
3	−0.408955910

4.6 Inclusion of Monoexcitations

In this Section we show how coupled cluster equations described in Chap. 3 are modified when monoexcitations are included. We thus obtain what is called CCSD, coupled clusters with singles and doubles. In this approximation we take cluster operator in the form

$$\hat{T} \simeq \hat{T}_1 + \hat{T}_2, \qquad (4.31)$$

where the one- and two-body cluster operators are given by Eqs. (3.28) and (3.29), $\hat{T}_1 = t_a^r \hat{e}_a^r$ and $\hat{T}_2 = \frac{1}{2!^2} t_{ab}^{rs} \hat{e}_{ab}^{rs}$. In the basis (3.2) the matrix elements of the cluster operator $\hat{T}$ are

$$\hat{T} = \begin{pmatrix} 0 & 0 & 0 & 0 & 0 \\ T_{20} & 0 & 0 & 0 & 0 \\ T_{40} & T_{42} & 0 & 0 & 0 \\ 0 & T_{62} & T_{64} & 0 & 0 \\ 0 & 0 & T_{84} & T_{86} & 0 \end{pmatrix},$$

so the Schrödinger equation (3.23) in the basis (3.2) reads

$$\begin{pmatrix} 1 & 0 & 0 \\ V_{20}^{-1} & 1 & 0 \\ V_{40}^{-1} & V_{42}^{-1} & 1 \end{pmatrix} \begin{pmatrix} 0 & 0 & H_{04} & 0 & 0 \\ 0 & H_{22} & H_{24} & H_{26} & 0 \\ H_{40} & H_{42} & H_{44} & H_{46} & H_{48} \end{pmatrix} \begin{pmatrix} 1 \\ V_{20} \\ V_{40} \\ V_{60} \\ V_{80} \end{pmatrix} = E_{\text{corr}} \begin{pmatrix} 1 \\ 0 \\ 0 \end{pmatrix}, \qquad (4.32)$$

where

$$V_{20} = T_{20}, \ V_{20}^{-1} = -T_{20}, \ V_{40} = T_{40} + \frac{1}{2} T_{42} T_{20}, \ V_{40}^{-1} = -T_{40} + \frac{1}{2} T_{42} T_{20},$$
(4.33)
$$V_{42}^{-1} = -T_{42}, \ V_{60} = \frac{1}{2}(T_{62} T_{20} + T_{64} T_{40}) + \frac{1}{3!} T_{64} T_{42} T_{20},$$

$$V_{80} = \frac{1}{2} T_{84} T_{40} + \frac{1}{3!}[T_{84} T_{42} T_{20} + T_{86}(T_{62} T_{20} + T_{64} T_{40})] + \frac{1}{4!} T_{86} T_{64} T_{42} T_{20}.$$

Equations (4.32) represents system of three equations. The first equation yields the relation between the correlation energy and cluster amplitudes,

$$E_{\text{corr}} = H_{04} \left(T_{40} + \frac{1}{2} T_{42} T_{20} \right),$$

and the second and third equations yield coupled equations for the cluster amplitudes

$$0 = [H_{24}T_{40}] + \left[H_{22}T_{20} - T_{20}H_{04}T_{40} + \frac{1}{2}H_{26}(T_{62}T_{20} + T_{64}T_{40})\right] +$$

$$+ \left[\frac{1}{2}H_{24}T_{42}T_{20}\right] + \left[\left(-\frac{1}{2}T_{20}H_{04} + \frac{1}{3!}H_{26}T_{64}\right)T_{42}T_{20}\right]$$

and

$$0 = \left[H_{40} + H_{42}T_{20} + \frac{1}{2}H_{44}T_{42}T_{20} - T_{42}H_{22}T_{20}\times\right.$$

$$\times \left(-\frac{1}{2}T_{42}H_{24} + \frac{1}{3!}H_{46}T_{64}\right)T_{42}T_{20}+$$

$$+ \left(\frac{1}{4}T_{42}T_{20}H_{04} - \frac{1}{3!}T_{42}H_{26}T_{64} + \frac{1}{4!}H_{48}T_{86}T_{64}\right)T_{42}T_{20}\right] +$$

$$+ \left[H_{44}T_{40} + \frac{1}{2}H_{46}(T_{62}T_{20} + T_{64}T_{40}) - T_{42}H_{24}T_{40}+\right.$$

$$+ \frac{1}{3!}H_{48}(T_{84}T_{42}T_{20} + T_{86}T_{62}T_{20}+$$

$$+ T_{86}T_{64}T_{40}) + \frac{1}{2}(T_{42}T_{20}H_{04}T_{40} - T_{40}H_{04}T_{42}T_{20}-$$

$$- T_{42}H_{26}(T_{62}T_{20} + T_{64}T_{40}))\right] +$$

$$+ \left[\frac{1}{2}H_{48}T_{84}T_{40} - T_{40}H_{04}T_{40}\right].$$

Going now to the explicit evaluations we have correspondence

$$\frac{1}{2}H_{04}T_{42}T_{20} \leftrightarrow \frac{1}{2}\langle F| : \hat{H} : \hat{T}_1^2|F\rangle = \frac{1}{2}v_{ab}^{rs}t_a^r t_b^s,$$

$$H_{24}T_{40} \leftrightarrow \langle F|\hat{e}_r^a : \hat{H} : \hat{T}_2|F\rangle = \frac{t_{cd}^{uv}}{4}\left[\mathcal{A}_{cd}v_{rd}^{uv}\delta_c^a - \mathcal{A}^{uv}v_{cd}^{ua}\delta_r^v\right],$$

$$H_{22}T_{20} \leftrightarrow \langle F|\hat{e}_r^a : \hat{H} : \hat{T}_1|F\rangle = t_b^s\left[v_{br}^{sa} + (\epsilon_s - \epsilon_b)\delta_r^s\delta_b^a\right],$$

4.6 Inclusion of Monoexcitations

$$-T_{20}H_{04}T_{40} + \frac{1}{2}H_{26}(T_{62}T_{20} + T_{64}T_{40})$$

$$\leftrightarrow \langle F|\hat{e}_r^a \left[-\hat{T}_1 : \hat{H} : \hat{T}_2 + \frac{1}{2} : \hat{H} : (\hat{T}_2\hat{T}_1 + \hat{T}_1\hat{T}_2)\right]|F\rangle =$$

$$= \frac{t_{cd}^{uv}t_b^s}{4}\left[\mathcal{A}^{uv}\delta_b^a\delta_r^u v_{cd}^{vs} + \mathcal{A}_{cd}\delta_d^a\delta_r^s v_{bc}^{uv} + \mathcal{A}_{cd}^{uv}\delta_d^a\delta_r^u v_{bc}^{vs}\right],$$

$$\frac{1}{2}H_{24}T_{42}T_{20} \leftrightarrow \frac{1}{2}\langle F|\hat{e}_r^a : \hat{H} : \hat{T}_1^2|F\rangle = t_c^u t_d^v \left[v_{rd}^{uv}\delta_c^a - v_{cd}^{ua}\delta_r^v\right],$$

$$\left(-\frac{1}{2}T_{20}H_{04} + \frac{1}{3!}H_{26}T_{64}\right)T_{42}T_{20}$$

$$\leftrightarrow \langle F|\hat{e}_r^a\left(-\frac{1}{2}\hat{T}_1 : \hat{H} : + \frac{1}{3!} : \hat{H} : \hat{T}_1\right)\hat{T}_1^2|F\rangle = t_b^s t_e^x t_f^y \delta_r^x \delta_b^a v_{ef}^{ys},$$

$$H_{42}T_{20} \leftrightarrow \langle F|\hat{e}_{rs}^{ab} : \hat{H} : \hat{T}_1|F\rangle = t_c^u\left[\mathcal{A}^{ab}v_{rs}^{ub}\delta_c^a + \mathcal{A}_{rs}v_{rc}^{ab}\delta_s^u\right]$$

$$\left(\frac{1}{2}H_{44}T_{42} - T_{42}H_{22}\right)T_{20} \leftrightarrow \langle F|\hat{e}_{rs}^{ab}\left(\frac{1}{2} : \hat{H} : \hat{T}_1 - \hat{T}_1 : \hat{H} :\right)\hat{T}_1|F\rangle =$$

$$= \frac{t_c^u t_d^v}{2}\left[\mathcal{A}^{ab}v_{rs}^{uv}\delta_c^a\delta_d^b + \mathcal{A}_{rs}v_{cd}^{ab}\delta_r^u\delta_s^v + 2\mathcal{A}_{rs}^{ab}v_{sd}^{ub}\delta_c^a\delta_r^v\right],$$

$$\left(-\frac{1}{2}T_{42}H_{24} + \frac{1}{3!}H_{46}T_{64}\right)T_{42}T_{20}$$

$$\leftrightarrow \langle F|\hat{e}_{rs}^{ab}\left(-\frac{1}{2}\hat{T}_1 : \hat{H} : + \frac{1}{3!} : \hat{H} : \hat{T}_1\right)\hat{T}_1^2|F\rangle =$$

$$= \mathcal{A}_{rs}^{ab}\frac{t_e^x t_c^u t_d^v}{2}\left[v_{er}^{uv}\delta_c^a\delta_d^b\delta_s^x + v_{cd}^{xb}\delta_e^a\delta_r^u\delta_s^v\right],$$

$$\left(\frac{1}{4}T_{42}T_{20}H_{04} - \frac{1}{3!}T_{42}H_{26}T_{64} + \frac{1}{4!}H_{48}T_{86}T_{64}\right)T_{42}T_{20} \leftrightarrow$$

$$\leftrightarrow \langle F|\hat{e}_{rs}^{ab}\left(\frac{1}{4}\hat{T}_1^2 : \hat{H} : -\frac{1}{3!}\hat{T}_1 : \hat{H} : \hat{T}_1 + \frac{1}{4!} : \hat{H}\hat{T}_1^2\right)\hat{T}_1^2|F\rangle = \frac{t_c^u t_d^v t_e^x t_f^y}{4} v_{cd}^{xy}\Delta_{ef}^{ab}\Delta_{rs}^{uv},$$

$$\frac{1}{2}H_{46}(T_{62}T_{20} + T_{64}T_{40}) - T_{42}H_{24}T_{40}$$

$$\leftrightarrow \langle F|\hat{e}^{ab}_{rs}\left[\frac{1}{2}:\hat{H}:\left(\hat{T}_2\hat{T}_1+\hat{T}_1\hat{T}_2\right)-\hat{T}_1:\hat{H}:\hat{T}_2\right]|F\rangle =$$

$$= \frac{t^{uv}_{cd}t^{x}_{e}}{4}\mathcal{A}^{ab}_{rs}\left[\delta^a_c\delta^b_d\left(\delta^x_s v^{uv}_{er}-\mathcal{A}^{uv}\delta^u_s v^{xv}_{er}\right)+\delta^u_r\delta^v_s\left(\delta^a_e v^{xb}_{cd}-\mathcal{A}_{cd}\delta^a_c v^{xb}_{ed}\right)+$$

$$+\mathcal{A}^{uv}_{cd}\delta^a_c\left(\delta^b_e\delta^u_s v^{xv}_{dr}+\delta^x_s\delta^u_r v^{vb}_{ed}\right)\right]$$

and finally

$$\frac{1}{3!}H_{48}\left(T_{84}T_{42}T_{20}+T_{86}T_{62}T_{20}+T_{86}T_{64}T_{40}\right)+$$

$$+\frac{1}{2}\left(T_{42}T_{20}H_{04}T_{40}-T_{40}H_{04}T_{42}T_{20}-T_{42}H_{26}\left(T_{62}T_{20}+T_{64}T_{40}\right)\right)\leftrightarrow$$

$$\leftrightarrow \langle F|\hat{e}^{ab}_{rs}\left[\frac{1}{3!}:\hat{H}:\left(\hat{T}_2\hat{T}_1^2+\hat{T}_1\hat{T}_2\hat{T}_1+\hat{T}_1^2\hat{T}_2\right)+\right.$$

$$\left.+\frac{1}{2}\left(\hat{T}_1^2:\hat{H}:\hat{T}_2-\hat{T}_2:\hat{H}:\hat{T}_1^2-\hat{T}_1:\hat{H}:\left(\hat{T}_2\hat{T}_1+\hat{T}_1\hat{T}_2\right)\right)\right]|F\rangle =$$

$$=\frac{t^{uv}_{cd}t^x_e t^y_f}{2^3}\mathcal{A}^{ab}_{rs}\left[v^{uv}_{ef}\delta^x_r\delta^y_s\delta^a_c\delta^b_d+v^{xy}_{cd}\delta^u_r\delta^v_s\delta^a_e\delta^b_f-\right.$$

$$\left.-2\mathcal{A}_{cd}v^{xy}_{fd}\delta^u_r\delta^v_s\delta^a_e\delta^b_c-2\mathcal{A}^{uv}v^{ux}_{ef}\delta^a_c\delta^b_d\delta^v_r\delta^y_s-2\mathcal{A}^{uv}_{cd}v^{uy}_{ed}\delta^a_f\delta^b_c\delta^x_r\delta^v_s\right].$$

From the cluster amplitudes t^r_a we can form symmetry adapted amplitudes

$$t^R_A(j) = \sum_{m_a,m_r}(j_r,m_r,j_a,-m_a|j)(-1)^{j_a-m_a}t^r_a. \qquad (4.34)$$

The cluster amplitudes with different j again decouple and only the amplitude with $j = 0$ contributes to the ground state energy. It follows from the properties of CG coefficients that

$$(j_r,m_r,j_a,-m_a|0)(-1)^{j_a-m_a} = \frac{\delta^{j_r}_{j_a}\delta^{m_r}_{m_a}}{\sqrt{2j_a+1}},$$

so Eq. (4.34) simplifies to

$$t^R_A = t^R_A(0)\sqrt{2j_a+1} = \delta^{j_r}_{j_a}\delta^{m_r}_{m_a}t^r_a.$$

Thus, the symmetry adaptation of the one-body cluster amplitudes is very simple, the quantum numbers of virtual spin-orbital have to be the same as occupied spin-orbital.

For the Hubbard model of benzene, inclusion of the one-body clusters has no effect on the correlation energy. As already mentioned in the previous chapter, in case of stable HF solution their effect is small. Nonetheless, as there is a small number of them, it is simple to take them into account, as it is usually done.

4.7 Perturbative Inclusion of Triexcitations

As also mentioned in the previous chapter, the triexcited configurations are so numerous, that their contribution is usually estimated only perturbatively. Here, we describe how it can be done.

We neglect contribution of the two-particle states and assume the Fermi vacuum, four-, six- and eight-particle states form an approximately complete basis set, cf. Eq. (3.31),

$$\hat{1} \simeq |0\rangle\langle 0| + |4\rangle\langle 4| + |6\rangle\langle 6| + |8\rangle\langle 8|. \tag{4.35}$$

The cluster operator is assumed to be of the form

$$\hat{T} \simeq \hat{T}_2 + \hat{T}_3.$$

In the basis (4.35) its matrix elements read, cf. Eq. (3.33),

$$\hat{T} = \begin{pmatrix} 0 & 0 & 0 & 0 \\ T_{40} & 0 & 0 & 0 \\ T_{60} & 0 & 0 & 0 \\ 0 & T_{84} & 0 & 0 \end{pmatrix}.$$

We then proceed in the same manner as in Sect. 3.3. We obtain from Eq. (3.23) the three equations, cf. Eqs. (3.34) and (3.35),

$$H_{04}T_{40} = E_{\text{corr}}, \tag{4.36}$$

$$H_{40} + H_{44}T_{40} - T_{40}H_{04}T_{40} + H_{46}T_{60} + \frac{1}{2}H_{48}T_{84}T_{40} = 0 \tag{4.37}$$

and

$$-T_{60}H_{04}T_{40} + H_{64}T_{40} + H_{66}T_{60} + \frac{1}{2}H_{68}T_{84}T_{40} = 0. \tag{4.38}$$

The last equation is an approximate one. Clearly, we would need to include more than eight-particle states to be exact, but it suffices for perturbative estimate of contribution of three-body clusters. We make formal expansion of the cluster operators

$$\hat{T}_2 = \hat{T}_2^{(0)} + \hat{T}_2^{(2)} + \ldots, \quad \hat{T}_3 = \hat{T}_3^{(1)} + \ldots$$

and the correlation energy

$$E_{\text{corr}} = E_{\text{corr}}^{(0)} + E_{\text{corr}}^{(2)} + \ldots,$$

where $E_{\text{corr}}^{(0)}$ and $\hat{T}_2^{(0)}$ are determined from Eqs. (3.34) and (3.35). Further, we split the matrix elements of Hamiltonian into one-electron and two-electron parts,

$$H_{44} = H_{44}^{(1)} + H_{44}^{(2)}, \quad H_{66} = H_{66}^{(1)} + H_{66}^{(2)}.$$

Then we get from Eq. (4.38)

$$H_{66}^{(1)} T_{60}^{(1)} = -H_{64} T_{40}^{(0)},$$

from Eq. (4.37)

$$H_{44}^{(1)} T_{40}^{(2)} = -H_{46} T_{60}^{(1)}$$

and from Eq. (4.36)

$$E_{\text{corr}}^{(2)} = H_{04} T_{40}^{(2)}.$$

Going now to the explicit evaluation we have correspondence, cf. Eq. (3.2),

$$H_{46} \leftrightarrow \frac{1}{3!^2} \langle F | \hat{e}_{rs}^{ab} : \hat{H} : \hat{e}_{cde}^{tuv} | F \rangle = \frac{1}{3!^2 4} v_{\mu\nu}^{\kappa\lambda} \langle F | \hat{e}_{rs}^{ab} : \hat{e}_{\kappa\lambda}^{\mu\nu} : \hat{e}_{cde}^{tuv} | F \rangle =$$

$$= \frac{1}{3!^2 4} v_{\mu\nu}^{\kappa\lambda} \left(\mathcal{A}_{rs}^{\mu\nu} \delta_r^\nu \Delta_{cde}^{ab\mu} \Delta_{\kappa\lambda s}^{tuv} + \mathcal{A}_{\kappa\lambda}^{ab} \delta_\lambda^b \Delta_{cde}^{a\mu\nu} \Delta_{\kappa rs}^{tuv} \right) =$$

$$= \frac{1}{3!^2} \mathcal{A}_{cde}^{tuv} \left[\mathcal{A}_{rs} \delta_s^\nu \Delta_{cd}^{ab} v_{er}^{tu} + \mathcal{A}^{ab} \delta_c^a \Delta_{rs}^{uv} v_{de}^{tb} \right].$$

Here the symbol $\mathcal{A}_{cde}$ denotes even permutations of indices c, d and e. In the last equality we used, for instance,

$$\Delta_{cde}^{ab\mu} = \langle F | \hat{b}_a^+ \hat{b}_b^+ \hat{b}_\mu^+ \hat{b}_e \hat{b}_d \hat{b}_c | F \rangle = \mathcal{A}_{cde} \delta_e^\mu \Delta_{cd}^{ab}.$$

4.7 Perturbative Inclusion of Triexcitations

Since Hamiltonian is Hermitian, the matrix elements H_{64} are obtained from relation

$$H_{64} \leftrightarrow \frac{1}{2^2}\left(\langle F|\hat{e}^{ab}_{rs}:\hat{H}:\hat{e}^{tuv}_{cde}|F\rangle\right)^*.$$

Further, the one-electron matrix elements are, cf. Eq. (3.47),

$$H^{(1)}_{44} \leftrightarrow \frac{1}{2^2}\epsilon_\mu\langle F|\hat{e}^{ab}_{rs}:\hat{e}^\mu_\mu:\hat{e}^{pq}_{ef}|F\rangle = \frac{1}{2^2}(\epsilon_r+\epsilon_s-\epsilon_a-\epsilon_b)\Delta^{ab}_{ef}\Delta^{pq}_{rs}$$

and

$$H^{(1)}_{66} \leftrightarrow \frac{1}{3!^2}\epsilon_\mu\langle F|\hat{e}^{cde}_{tuv}:\hat{e}^\mu_\mu:\hat{e}^{xyz}_{fgh}|F\rangle$$

$$= \frac{1}{3!^2}(\epsilon_t+\epsilon_u+\epsilon_v-\epsilon_c-\epsilon_d-\epsilon_e)\Delta^{cde}_{fgh}\Delta^{xyz}_{tuv}.$$

Symmetry adaptation of the three-body cluster amplitudes is accomplished as follows. We first make transition from unordered to ordered amplitudes, cf. Eq. (4.4),

$$t^{qrs}_{abc} = \frac{1}{3!}\mathcal{P}^{qrs}_{abc}\operatorname{sgn}(\mathcal{P}^{qrs})\operatorname{sgn}(\mathcal{P}_{abc})T^{\overline{qrs}}_{\overline{abc}},$$

where $\mathcal{P}$ stands for all permutations of the indices involved. Further, we form linear combinations, cf. Eq. (4.13),

$$T^{\overline{qrs}}_{\overline{abc}} = \sum_{J_{rs},J_{bc},J} D^{j_q,m_q,j_r,m_r,j_s,m_s,\delta_{QR},\delta_{RS}}_{j_a,m_a,j_b,m_b,j_c,m_c,\delta_{AB},\delta_{BC}}(J_{bc},J_{rs},J)T^{\overline{QRS}}_{\overline{ABC}}(J_{bc},J_{rs},J)$$

(4.39)

to create amplitudes with total zero spin. Here the D coefficients are given by the product

$$D^{j_q,m_q,j_r,m_r,j_s,m_s,\delta_{QR},\delta_{RS}}_{j_a,m_a,j_b,m_b,j_c,m_c,\delta_{AB},\delta_{BC}}(J_{bc},J_{rs},J) = V^{j_q,j_r,j_s,\delta_{QR},\delta_{RS}}_{j_a,j_b,j_c,\delta_{AB},\delta_{BC}}(J_{bc},J_{rs},J)\delta^{m_q+m_r+m_s}_{m_a+m_b+m_c}\times$$

(4.40)

$$\times G_{j_q,m_q,j_r,m_r,j_s,m_s,\delta_{QR},\delta_{RS}}(J_{rs},J)G_{j_a,m_a,j_b,m_b,j_c,m_c,\delta_{AB},\delta_{BC}}(J_{bc},J),$$

where

$$G_{j_q,m_q,j_r,m_r,j_s,m_s,\delta_{QR},\delta_{RS}}(J_{rs},J) =$$
$$= (1-\delta_{QR})(j_r,m_r,j_s,m_s|J_{rs})(j_q,m_q,J_{rs},m_r+m_s|J_+)\times$$

(4.41)

$$\times\left\{1-\delta_{RS}+\delta_{RS}\Theta(m_s-m_r)\left[1-(-1)^{J_{rs}-j_r-j_s}\right]\right\}+$$

$$+ \delta_{QR}\delta_{RS}\Theta(m_s - m_r)\Theta(m_r - m_q) \times$$

$$\times \left[\delta_{J_{rs}, j_r + j_s} \frac{1 + (-1)^{j_r + j_s}}{2} + \delta_{J_{rs}, j_r + j_s - 1} \frac{1 - (-1)^{j_r + j_s}}{2} \right] \times$$

$$\times \sum_{i=1}^{6} \mathrm{sgn}(\mathcal{P}_i)(j_{\mathcal{P}_i(r)}, m_{\mathcal{P}_i(r)}, j_{\mathcal{P}_i(s)}, m_{\mathcal{P}_i(s)} | J_{rs}) \times$$

$$\times (j_{\mathcal{P}_i(q)}, m_{\mathcal{P}_i(q)}, J_{rs}, m_{\mathcal{P}_i(r)} + m_{\mathcal{P}_i(s)} | J) \}.$$

V coefficients in Eq. (4.40) are there to ensure proper normalization, cf. Eq. (4.15). The last equation requires some explanation. In general we have to distinguish three cases concerning three virtual or occupied orbitals: (1) all of them are different, (2) the two of them are the same and the third is different, (3) all of them are the same. Since the orbitals are ordered, the second case can be always arranged in such a way that the same orbitals are labeled by R and S. In the third case the value of J_{rs} is irrelevant, as all the possible symmetry adapted states with different values of J_{rs} are linearly dependent. For definitness we choose the highest possible value. In the third case the sum over all possible permutations is there to exclude the values of J_{rs} and J that do not contribute because Pauli exclusion principle is violated.

If we include the contribution of the three-body clusters to the correlation energy of Hubbard model of benzene in the manner described above we get $E_{\mathrm{corr}}/t \simeq -0.409688$. This slightly improves agreement of the coupled cluster result with the exact solution $E_{\mathrm{corr}}/t \simeq -0.40944$.

4.8 One-electron Open Shells

4.8.1 Combination of Coupled Clusters and Configuration Interaction

For one-electron open-shell atoms and molecules one can use the coupled cluster result for closed-shell systems as follows.[3] Firstly, we solve Hartree-Fock equations for pertinent closed-shell cation. This yields Fermi vacuum state, Eqs. (2.10), (2.11) and (2.12), Secondly, we assume the valence electron moves in the potential created by core electrons. This provides us with an one-electron basis set. As already alluded to in Sect. 2.5 the advantage of this procedure is that solution of Hartree-Fock equations for pertinent closed-shell cation preserves the spin symmetry of the

---

[3] The method is usually referred to as equation of motion coupled cluster method, see [4] and references therein, rather unappealing and misleading name, there is no time evolution and hence no equations of motion to be solved. The method firmly states in the framework of quantum electrostatics, so to speak.

4.8 One-electron Open Shells

molecular Hamiltonian (3.38). Thirdly, instead of Eq. (3.22), we search for solution of Schrödinger equation, cf. Eq. (3.7),

$$:\hat{H}: |\psi\rangle = E_{\text{corr,os}}|\psi\rangle$$

in the form

$$|\psi\rangle = \exp\{\hat{T}\}|\varphi\rangle, \tag{4.42}$$

where $\hat{T} \simeq \hat{T}_2$ is the operator determined previously for the closed-shell cation, see Eqs. (3.39) and (3.40). Substituting the last equation into the penultimate equation and multiplying the resulting equation by $\exp\{-\hat{T}\}$ from the left we obtain

$$\overline{\hat{H}}|\varphi\rangle = E_{\text{corr,os}}|\varphi\rangle, \quad \overline{\hat{H}} = \exp\{-\hat{T}\} : \hat{H} : \exp\{\hat{T}\}. \tag{4.43}$$

Further, we assume one, three, five and seven particle states form an approximately complete set,

$$\hat{1} \simeq |1\rangle\langle 1| + |3\rangle\langle 3| + |5\rangle\langle 5| + |7\rangle\langle 7|, \tag{4.44}$$

where in analogy with Eq. (3.2) we introduced the following correspondence for the excitations from the Fermi vacuum

$$|1\rangle \leftrightarrow \hat{e}^r|F\rangle, \quad \langle 1| \leftrightarrow \langle F|\hat{e}_r, \quad |3\rangle \leftrightarrow \frac{1}{2!}\hat{e}^{rs}_a|F\rangle, \quad \langle 3| \leftrightarrow \langle F|\hat{e}^a_{rs}, \tag{4.45}$$

$$|5\rangle \leftrightarrow \frac{1}{3!2!}\hat{e}^{rst}_{ab}|F\rangle, \quad \langle 5| \leftrightarrow \langle F|\hat{e}^{ab}_{rst}$$

and so on. In this basis Eq. (4.42) takes the form

$$\begin{pmatrix} c_1 \\ c_3 \\ c_5 \\ c_7 \end{pmatrix} = \begin{pmatrix} 1 & 0 & 0 & 0 \\ 0 & 1 & 0 & 0 \\ T_{51} & 0 & 1 & 0 \\ 0 & T_{73} & 0 & 1 \end{pmatrix} \begin{pmatrix} f_1 \\ f_3 \\ f_5 \\ f_7 \end{pmatrix}.$$

Further, Eq. (4.43) projected on the one- and three-particle states takes the form

$$\begin{pmatrix} \overline{H}_{11} & \overline{H}_{13} \\ \overline{H}_{31} & \overline{H}_{33} \end{pmatrix} \begin{pmatrix} f_1 \\ f_3 \end{pmatrix} = E_{\text{corr,os}} \begin{pmatrix} f_1 \\ f_3 \end{pmatrix}, \tag{4.46}$$

where

$$\begin{pmatrix} \overline{H}_{11} & \overline{H}_{13} \\ \overline{H}_{31} & \overline{H}_{33} \end{pmatrix} = \begin{pmatrix} 1 & 0 & 0 & 0 \\ 0 & 1 & 0 & 0 \end{pmatrix} \times \quad (4.47)$$

$$\times \begin{pmatrix} H_{11} & H_{13} & H_{15} & 0 \\ H_{31} & H_{33} & H_{35} & H_{37} \\ H_{51} & H_{53} & H_{55} & H_{57} \\ 0 & H_{73} & H_{75} & H_{77} \end{pmatrix} \begin{pmatrix} 1 & 0 \\ 0 & 1 \\ T_{51} & 0 \\ 0 & T_{73} \end{pmatrix}.$$

One can see the idea here is similar to the closed shell case. Though we end up with configuration interaction involving, say, one- and three-particle configurations, the contribution of five- and seven-particle configurations is somehow taken into account as well. Again note that had we restricted ourselves to the one- and three-particle configurations from the very beginning, Eq. (4.46) would yield the same correlation energies as configuration interactions with one- and three-particle configurations,

$$\begin{pmatrix} H_{11} & H_{13} \\ H_{31} & H_{33} \end{pmatrix} \begin{pmatrix} c_1 \\ c_3 \end{pmatrix} = E_{\text{corr,os}} \begin{pmatrix} c_1 \\ c_3 \end{pmatrix}. \quad (4.48)$$

4.8.2 Method for Obtaining Bound-state Energies

For calculation of bound-state energies one does not need to solve full Eq. (4.46). The three- and even majority of one-particle configurations lay in continuum. Thus we split the space of states on two parts, A and B. The part A comprises states for which $\overline{H}_{ii} < 0$. The part B comprises all other states. Equation (4.46) then can be rewritten as

$$\begin{pmatrix} \overline{H}_{AA} & \overline{H}_{AB} \\ \overline{H}_{BA} & \overline{H}_{BB} \end{pmatrix} \begin{pmatrix} f_A \\ f_B \end{pmatrix} = E_{\text{corr,os}} \begin{pmatrix} f_A \\ f_B \end{pmatrix}. \quad (4.49)$$

Now, we make another similarity transformation

$$\begin{pmatrix} f_A \\ f_B \end{pmatrix} = \begin{pmatrix} 1 & 0 \\ S_{BA} & 1 \end{pmatrix} \begin{pmatrix} d_A \\ 0 \end{pmatrix}.$$

Substituting this Ansatz into Eq. (4.49) we obtain

$$\begin{pmatrix} 1 & 0 \\ -S_{BA} & 1 \end{pmatrix} \begin{pmatrix} \overline{H}_{AA} & \overline{H}_{AB} \\ \overline{H}_{BA} & \overline{H}_{BB} \end{pmatrix} \begin{pmatrix} 1 \\ S_{BA} \end{pmatrix} d_A = E_{\text{corr,os}} \begin{pmatrix} d_A \\ 0 \end{pmatrix}. \quad (4.50)$$

4.8 One-electron Open Shells

The last equation represents a system of two equations. The second equation is equation for matrix elements S_{BA}

$$\overline{H}_{BA} + \overline{H}_{BB}S_{BA} - S_{BA}\overline{H}_{AA} - S_{BA}\overline{H}_{AB}S_{BA} = 0 \tag{4.51}$$

The first equation is the equation for the correlation energy,

$$\mathcal{H}_{AA}d_A = E_{\text{corr,os}}d_A, \tag{4.52}$$

where the effective one-particle Hamiltonian $\mathcal{H}_{AA}$ is given by

$$\mathcal{H}_{AA} = \overline{H}_{AA} + \overline{H}_{AB}S_{BA}. \tag{4.53}$$

Equation (4.51) is solved iteratively. Separating in matrices H_{AA} and H_{BB} the diagonal (D) and off-diagonal (W) parts

$$\overline{H}_{AA} = D_{AA} + W_{AA} \qquad \overline{H}_{BB} = D_{BB} + W_{BB}$$

one writes in the l-th iteration

$$\left(S_{BA}^{(l)}\right)_{ij} = \frac{1}{(D_{BB})_{ii} - (D_{AA})_{jj}} \times \tag{4.54}$$

$$\times \left[-\overline{H}_{BA} - W_{BB}S_{BA}^{(l-1)} + S_{BA}^{(l-1)}W_{AA} + S_{BA}^{(l-1)}\overline{H}_{AB}S_{BA}^{(l-1)}\right]_{ij}.$$

The iteration is initialized by setting $S_{BA}^{(-1)} = 0$. The method works if the diagonal matrix elements are considerably larger than off-diagonal matrix elements, $|D_{BB}| \gg |W_{BB}|$, and differences of diagonal matrix elements of B and A parts, $|(D_{BB})_{ii} - (D_{AA})_{jj}|$, sufficiently large.

4.8.3 Matrix Elements of Configuration Interaction

Performing the matrix multiplication on rhs of Eq. (4.47) yields

$$\overline{H}_{11} = H_{11} + H_{15}T_{51}, \; \overline{H}_{31} = H_{31} + H_{35}T_{51}, \; \overline{H}_{13} = H_{13}, \; \overline{H}_{33} = H_{33} + H_{37}T_{73}. \tag{4.55}$$

Going now to the explicit evaluation using the correspondence (4.45) we get

$$H_{11} \leftrightarrow \langle F | \hat{e}_q : \hat{H} : \hat{e}^p | F \rangle = \epsilon_q \delta_q^p, \tag{4.56}$$

$$H_{15}T_{51} \leftrightarrow \langle F|\hat{e}_q : \hat{H} : \hat{T}_2 \hat{e}^p|F\rangle = \frac{1}{4^2} v_{\mu\nu}^{\kappa\lambda} t_{cd}^{uv} \langle F|\hat{e}^q : \hat{e}_{\kappa\lambda}^{\mu\nu} : \hat{e}_{cd}^{uv} \hat{e}^p|F\rangle = \quad (4.57)$$

$$= \frac{1}{4^2} v_{\mu\nu}^{\kappa\lambda} t_{cd}^{uv} \Delta_{cd}^{\mu\nu} \left(\delta_q^p \Delta_{\kappa\lambda}^{uv} - \mathcal{A}^{uv} \delta_q^v \Delta_{\kappa\lambda}^{up}\right) = \frac{t_{cd}^{uv}}{4} \left(\delta_q^p v_{cd}^{uv} - \mathcal{A}^{uv} \delta_q^v v_{cd}^{up}\right),$$

$$H_{31} \leftrightarrow \langle F|\hat{e}_{qr}^a : \hat{H} : \hat{e}^p|F\rangle =$$

$$= \frac{1}{4} v_{\kappa\lambda}^{\mu\nu} \langle F|\hat{e}_{qr}^a : \hat{e}_{\kappa\lambda}^{\mu\nu} : \hat{e}^p|F\rangle = \frac{1}{4} v_{\kappa\lambda}^{\mu\nu} \mathcal{A}^{\kappa\lambda} \delta_\lambda^a \delta_\kappa^p \Delta_{qr}^{\mu\nu} = v_{qr}^{pa}, \quad (4.58)$$

$$H_{35}T_{51} \leftrightarrow \langle F|\hat{e}_{qr}^a : \hat{H} : \hat{T}_2 \hat{e}^p|F\rangle = \frac{1}{4^2} v_{\kappa\lambda}^{\mu\nu} t_{cd}^{uv} \langle F|\hat{e}_{qr}^a : \hat{e}_{\kappa\lambda}^{\mu\nu} : \hat{e}_{cd}^{uv} \hat{e}^p|F\rangle =$$
(4.59)

$$= \frac{1}{4^2} v_{\kappa\lambda}^{\mu\nu} t_{cd}^{uv} \left[\mathcal{A}_{rq}^{\mu\nu} \Delta_{cd}^{a\mu} \delta_q^v \left(\delta_r^p \Delta_{\kappa\lambda}^{uv} - \mathcal{A}^{uv} \delta_r^v \Delta_{\kappa\lambda}^{up}\right) + \right.$$
$$\left. + \mathcal{A}_{\kappa\lambda} \Delta_{cd}^{\mu\nu} \delta_\lambda^a \left(\delta_\kappa^p \Delta_{rq}^{uv} - \mathcal{A}^{uv} \delta_\kappa^v \Delta_{rq}^{up}\right)\right] =$$

$$= \frac{t_{cd}^{uv}}{4} \left[\mathcal{A}_{rq} \mathcal{A}_{cd} \delta_c^a \left(v_{dq}^{uv} \delta_r^p - \mathcal{A}^{uv} \delta_r^v v_{dq}^{up}\right) + \Delta_{rq}^{uv} v_{cd}^{pa} - \mathcal{A}^{uv} \Delta_{rq}^{up} v_{cd}^{va}\right],$$

$$H_{33} \leftrightarrow \langle F|\hat{e}_{qr}^a : \hat{H} : \hat{e}_b^{pt}|F\rangle = \epsilon_\mu \langle F|\hat{e}_{qr}^a : \hat{e}_\mu^\mu : \hat{e}_b^{pt}|F\rangle +$$
$$+ \frac{1}{4} v_{\mu\nu}^{\kappa\lambda} \langle F|\hat{e}_{qr}^a : \hat{e}_{\kappa\lambda}^{\mu\nu} : \hat{e}_b^{pt}|F\rangle = \quad (4.60)$$

$$= \epsilon_\mu \left(\delta_\mu^a \delta_b^\mu \Delta_{qr}^{pt} + \mathcal{A}_{qr}^{pt} \delta_\mu^p \delta_q^\mu \delta_b^a\right) + \frac{1}{4} v_{\mu\nu}^{\kappa\lambda} \left(\delta_b^a \Delta_{\kappa\lambda}^{pt} \Delta_{qr}^{\mu\nu} + \mathcal{A}_{\kappa\lambda}^{\mu\nu} \mathcal{A}_{rq}^{pt} \delta_\lambda^a \delta_b^\mu \delta_\kappa^p \delta_q^v \delta_r^t\right) =$$

$$= (\epsilon_q + \epsilon_r - \epsilon_a) \delta_b^a \Delta_{qr}^{pt} + v_{qr}^{pt} \delta_b^a + \mathcal{A}_{qr}^{pt} \left\{v_{bq}^{pa} \delta_r^t\right\},$$

$$H_{37}T_{73} \leftrightarrow \frac{1}{2} \langle F|\hat{e}_{qr}^a : \hat{H} : \hat{T}_2 \hat{e}_b^{pt}|F\rangle = \frac{1}{2^5} v_{\mu\nu}^{\kappa\lambda} t_{cd}^{uv} \langle F|\hat{e}_{qr}^a : \hat{e}_{\kappa\lambda}^{\mu\nu} : \hat{e}_{cd}^{uv} \hat{e}_b^{pt}|F\rangle =$$
(4.61)
$$= \frac{1}{2^5} v_{\mu\nu}^{\kappa\lambda} t_{cd}^{uv} \left(\delta_b^a \Delta_{cd}^{\mu\nu} - \mathcal{A}_{cd} \delta_c^a \Delta_{bd}^{\mu\nu}\right) \left(\Delta_{\kappa\lambda}^{uv} \Delta_{qr}^{pt} + \Delta_{qr}^{uv} \Delta_{\kappa\lambda}^{pt} - \mathcal{A}^{uv} \mathcal{A}^{pt} \Delta_{\kappa\lambda}^{up} \Delta_{qr}^{vt}\right) =$$

$$= \frac{1}{2^3} t_{cd}^{uv} \left\{\delta_b^a \left[v_{cd}^{uv} \Delta_{qr}^{pt} + v_{cd}^{pt} \Delta_{qr}^{uv} - \mathcal{A}^{uv} \mathcal{A}^{pt} v_{cd}^{up} \Delta_{qr}^{vt}\right] - \right.$$
$$\left. - \mathcal{A}_{cd} \delta_c^a \left[v_{bd}^{uv} \Delta_{qr}^{pt} + v_{bd}^{pt} \Delta_{qr}^{uv} - \mathcal{A}^{uv} \mathcal{A}^{pt} v_{bd}^{up} \Delta_{qr}^{vt}\right]\right\},$$

4.8 One-electron Open Shells

Since Hamiltonian is Hermitian, the last matrix element H_{13} is obtained from relation

$$H_{13} \leftrightarrow \frac{1}{2} \left(\langle F | \hat{e}^a_{qr} : \hat{H} : \hat{e}^p | F \rangle \right)^*,$$

and Eq. (4.58). In Eq. (4.61) we used Eq. (3.50). In Eqs. (4.57), (4.59) and (4.61) we used the relation

$$\langle F | \hat{b}^+_a \hat{b}^+_\mu \hat{b}^+_\nu \hat{b}_d \hat{b}_c \hat{b}_b | F \rangle = \Delta^{a\mu\nu}_{bcd} = \delta^a_b \Delta^{\mu\nu}_{cd} - \mathcal{A}_{cd} \delta^a_c \Delta^{\mu\nu}_{bd}.$$

4.8.4 Perturbative Inclusion of Five-particle States

Similarly to the triexcitations in closed shell case, the five-particle states are so numerous that they are included in perturbative manner only. If we project Eq. (4.43) on five-particle states we obtain

$$\overline{H}_{51} = H_{51} + H_{55}T_{51} - T_{51}H_{11} - T_{51}H_{15}T_{51},$$

where

$$H_{51} \leftrightarrow \langle F | \hat{e}^{ab}_{rst} : \hat{H} : \hat{e}^p | F \rangle = \mathcal{A}_{rst} \left\{ \delta^q_t v^{ab}_{rs} \right\}, \qquad (4.62)$$

$$H_{55}T_{51} - T_{51}H_{11} \leftrightarrow \langle F | \hat{e}^{ab}_{rst} \left[: \hat{H} :, \hat{T}_2 \right] \hat{e}^p | F \rangle =$$

$$= \frac{t^{uv}_{cd}}{4} \left\{ \Delta^{uvp}_{rst} \Delta^{ab}_{cd} (\epsilon_r + \epsilon_s + \epsilon_t - \epsilon_a - \epsilon_b - \epsilon_p) + \right.$$

$$+ \mathcal{A}_{rst} \left[\Delta^{ab}_{cd} \left(v^{uv}_{rs} \delta^p_t - \mathcal{A}^{uv} v^{up}_{rs} \delta^v_t \right) + v^{ab}_{cd} \Delta^{uv}_{rs} \delta^p_t + \right.$$

$$+ \left. \mathcal{A}^{ab}_{cd} \delta^a_c \left(v^{bp}_{rd} \Delta^{uv}_{st} - \mathcal{A}^{uv} v^{bv}_{rd} \Delta^{up}_{st} \right) \right] \right\}$$

and

$$-T_{51}H_{15}T_{51} \leftrightarrow -\langle F | \hat{e}^{ab}_{rst} \hat{T}_2 : \hat{H} : \hat{T}_2 \hat{e}^p | F \rangle$$

$$= -\frac{t^{uv}_{cd} t^{xy}_{ef}}{4^2} \mathcal{A}_{rst} \left\{ \wedge^{ab}_{ef} \wedge^{xy}_{rs} (\delta^p_t v^{uv}_{cd} - \mathcal{A}^{uv} \delta^v_t v^{up}_{cd}) \right\}.$$

Further, the effective one-particle Hamiltonian, Eq. (4.53), is modified to

$$\mathcal{H}_{AA} = \overline{H}_{AA} + \overline{H}_{AB} S_{BA} + \overline{H}_{A5} S_{5A},$$

where $\overline{H}_{A5} = H_{A5}$ is determined from Eq. (4.62) and relation

$$H_{15} \leftrightarrow \frac{1}{2!3!} \left(\langle F | \hat{e}^{ab}_{rst} : \hat{H} : \hat{e}^p | F \rangle \right)^*.$$

Finally, S_{5A} is approximately given by, cf. Eq. (4.54),

$$S_{5A} \simeq -\frac{1}{\epsilon_r + \epsilon_s + \epsilon_t - \epsilon_a - \epsilon_b - \epsilon_A} \overline{H}_{5A}.$$

4.8.5 Symmetry Adaptation

Above equations can again be adapted to permutational and spin symmetry in the same spirit as for the closed shell case. We first make transition from unordered to ordered amplitudes

$$C^{qr}_a = \frac{1}{\sqrt{2}} \mathcal{A}^{qr} C^{\overline{qr}}_a, \quad C^{qrs}_{ab} = \frac{1}{\sqrt{3!2}} \mathcal{P}^{qrs}_{ab} \operatorname{sgn}(\mathcal{P}^{qrs}) \operatorname{sgn}(\mathcal{P}_{ab}) C^{\overline{qrs}}_{\overline{ab}}.$$

Further, we introduce the orthogonal transformations from the spin-orbital coefficients to the coefficients describing the states of definite square of the total spin J, projection to the one of the axes M and parity $(-1)^{J-\kappa/2}$, $\kappa = \pm 1$. These transformations read for one-particle

$$c^Q(J, M, \kappa) = c^q \delta^J_{j_q} \delta^M_{m_q} \delta^\kappa_{\kappa_q},$$

three-particle

$$c^{QR,J_+}_A(J, M, \kappa) = P^{(3)} \sum_{m_q, m_a} (j_q, m_q, j_r, M + m_a - m_q | J_+) \times$$

$$\times (J_+, M + m_a, j_a, -m_a | J)(-1)^{j_a - m_a} c^{qr}_a$$

and five-particle states

$$c^{QRS, J_{rs}, J_+}_{AB, J_-}(J, M, \kappa) = P^{(5)} \sum_{m_r, m_q, m_s, m_a, m_b} (j_r, m_r, j_s, m_s | J_{rs}) \times$$

$$\times (j_q, m_q, J_{rs}, m_r + m_s | J_+) \times$$

4.8 One-electron Open Shells

$$\times (j_a, m_a, j_b, m_b | J_-)(J_+, m_q + m_r + m_s, J_-, -m_a - m_b | J) \times$$

$$\times (-1)^{J_- - m_a - m_b} \delta^{m_q + m_r + m_s}_{M + m_a + m_b} c^{qrs}_{ab} ,$$

where the parity factors equal one for correct combination of particle and hole parities and otherwise vanish,

$$P^{(3)} = 1 \Leftrightarrow \left(j_q + j_r - j_a - \frac{\kappa_q + \kappa_r - \kappa_a}{2} \right) \mod 2 = \left(J - \frac{\kappa}{2} \right) \mod 2$$

and

$$P^{(5)} = 1 \Leftrightarrow \left(j_q + j_r + j_s - j_a - j_b - \frac{\kappa_q + \kappa_r + \kappa_s - \kappa_a - \kappa_b}{2} \right)$$

$$\mod 2 = \left(J - \frac{\kappa}{2} \right) \mod 2 .$$

The inverse tranformations for three-particle and five-particle states read

$$c^{qr}_a = P^{(3)} \sum_{J_+, J} (j_q, m_q, j_r, M + m_a - m_q | J_+) \times \qquad (4.63)$$

$$\times (J_+, M + m_a, j_a, -m_a | J)(-1)^{j_a - m_a} c^{QR, J_+}_A (J, M, \kappa)$$

and

$$c^{qrs}_{ab} = P^{(5)} \sum_{J_{rs}, J_+, J_-, J} (j_r, m_r, j_s, m_s | J_{rs})(j_q, m_q, J_{rs}, m_r + m_s | J_+) \times$$

$$\times (j_a, m_a, j_b, m_b | J_-) \times \qquad (4.64)$$

$$\times (J_+, m_q + m_r + m_s, J_-, -m_a - m_b | J) \times$$

$$\times (-1)^{J_- - m_a - m_b} \delta^{m_q + m_r + m_s}_{M + m_a + m_b} c^{QRS, J_{rs}, J_+}_{AB, J_-} (J, M, \kappa) ,$$

respectively. In the case of ordered amplitudes the transformations (4.63) and (4.64) are modified as follows,

$$C^{\overline{qr}}_a = P^{(3)} \sum_{J_+, J} D^{j_q, m_q, j_r, m_r, \delta_{QR}}_{j_a, m_a} (J_+, J) C^{\overline{QR}, J_+}_A (J, M, \kappa) , \qquad (4.65)$$

where

$$D^{j_q, m_q, j_r, m_r, \delta_{QR}}_{j_a, m_a} (J_+, J) = V^{j_q, j_r, \delta_{QR}}_{j_a} (J_+, J) \times \qquad (4.66)$$

$$\times (j_q, m_q, j_r, M + m_a - m_q | J_+)(J_+, M + m_a, j_a, -m_a | J)(-1)^{j_a - m_a} \times$$

$$\times \left\{ 1 - \delta_{QR} + \delta_{QR}\Theta(m_r - m_q)[1 - (-1)^{J_+ - j_r - j_q}] \right\}$$

and

$$C\frac{\overline{qrs}}{ab} = P^{(5)} \sum_{J_{rs}, J_+, J_-, J} D^{j_q, m_q, j_r, m_r, j_s, m_s, \delta_{QR}, \delta_{RS}}_{j_a, m_a, j_b, m_b, \delta_{AB}} \times$$

$$\times (J_{rs}, J_+, J_-, J) C\frac{\overline{QRS}, J_{rs}, J_+}{AB, J_-}(J, M, \kappa), \quad (4.67)$$

where

$$D^{j_q, m_q, j_r, m_r, j_s, m_s, \delta_{QR}, \delta_{RS}}_{j_a, m_a, j_b, m_b, \delta_{AB}}(J_{rs}, J_+, J_-, J) = V^{j_q, j_r, j_s, \delta_{QR}, \delta_{RS}}_{j_a, j_b, \delta_{AB}}(J_{rs}, J_+, J_-, J) \times \quad (4.68)$$

$$\times (J_+, m_q + m_r + m_s, J_-, -m_a - m_b | J)(-1)^{J_- - m_a - m_b} \delta^{m_q + m_r + m_s}_{M + m_a + m_b} \times$$

$$\times (j_a, m_a, j_b, m_b | J_-) \left\{ 1 - \delta_{AB} + \delta_{AB}\Theta(m_b - m_a)\left[1 - (-1)^{J_- - j_a - j_b}\right] \right\} \times$$

$$\times G_{j_q, m_q, j_r, m_r, j_s, m_s, \delta_{QR}, \delta_{RS}}(J_{rs}, J_+),$$

respectively. V's in Eqs. (4.66), (4.68) are there to ensure proper normalization, G's are given by Eq. (4.41).

As in the closed shell case, the configurations with different values of J and M decouple, so effectively only one value of J and M has to be considered on rhs of Eqs. (4.65) and (4.67).

References

1. A. Szabo, N.S. Ostlund, *Modern Quantum Chemistry. Introduction to Advanced Electronic Structure Theory* (Dover Publications, Mineola, 1996)
2. J. Čížek, J. Paldus, Int. J. Quantum Chem. **5**, 359 (1971)
3. J. Čížek, J. Chem. Phys. **45**, 4256 (1966); See also J. Čížek, Adv. Chem. Phys. **14**, 35 (1968); J. Paldus, J. Čížek, I. Shavitt, Phys. Rev. A **5**, 50 (1972); For historical perspective see R.J. Bartlett, Theor. Chem. Acc. **103**, 273 (2000)
4. R.J. Bartlett, M. Musial, Rev. Mod. Phys. **79**, 291 (2007)
5. J. Paldus, Nijmegen Lectures. Available at www.math.uwaterloo.ca/paldus/resources.html
6. I. Hubač, P. Čársky, in *Organic Chemistry and Theory. Topics in Current Chemistry*, vol. 75 (Springer, Berlin/Heidelberg, 1978)
7. I. Lindgren, J. Morrison, *Atomic Many-Body Theory* (Springer, Berlin, 1982)
8. J. Paldus, in *Relativistic and Electron Correlation Effects in Molecules and Solids, NATO ASI Series*, ed. by G.L. Malli (Plenum Press, New York, 1993)
9. J. Paldus, The beginnings of the coupled-cluster theory: an eyewitness account, in *Theory and Applications of Computational Chemistry: The First Forty Years*, ed. by C.E. Dykstra, G. Frenking, K.S. Kim, G.E. Scuseria, , Chap. 7 (Elsevier, Amsterdam, 2005), pp. 115–147
10. A. Messiah, *Quantum Mechanics* (North Holland Publishing, Amsterdam, 1961)
11. L. Schiff, *Quantum Mechanics* (McGraw Hill College, New York, 1968)
12. P.A.M. Dirac, *Principles of Quantum Mechanics* (Oxford University Press, Oxford, 1947)
13. L.D. Landau, E.M. Lifshitz, *Quantum Mechanics, Non-Relativistic Theory* (Elsevier/Butterworth-Heinemann, Amsterdam/Waltham, 1976)
14. J. Zamastil, J. Benda, *Quantum Mechanics and Electrodynamics* (Springer, Berlin, 2017)
15. D.J. Thouless, Nucl. Phys. **21**, 225 (1960); D.J. Thouless, *The Quantum Mechanics of Many-body Systems* (Academic Press, Cambridge, 1961)
16. J. Čížek, J. Paldus, J. Chem. Phys. **47**, 3976 (1967)
17. T. Uhlířová, J. Zamastil, Phys. Rev. A **101**, 062504 (2020)
18. P. Lykos, G.W. Pratt, Rev. Mod. Phys. **35**, 496 (1963)
19. R. Pauncz, *Spin Eigenfunctions: Construction and Use* (Springer, Berlin, 2012)
20. R.G. Parr, *The Quantum Theory of Molecular Electronic Structure*. A Lecture-note and Reprint Volume (W.A. Benjamin, New York, 1963)
21. R. Daudel, R. Lefebvre, C. Moser, *Quantum Chemistry. Methods and Applications* (Interscience Publishers, Inc., New York, 1965)
22. J. Čížek, F. Vinette, Chem. Phys. Lett. **149**, 516 (1988)

The manufacturer's authorised representative in the EU is Springer Nature Customer Service Centre GmbH, Europaplatz 3, 69115 Heidelberg, Germany. If you have any concerns regarding our products, please contact ProductSafety@springernature.com

Printed and bound by CPI Group (UK) Ltd, Croydon, CR0 4YY

26/03/2026

02078971-0002